BEAMS AND BEAM COLUMNS

Stability and Strength

Related titles

AXIALLY COMPRESSED STRUCTURES: STABILITY AND STRENGTH

edited by R. Narayanan

1. Centrally Compressed Members LAMBERT TALL
2. Current Trends in the Treatment of Safety I. H. G. DUNCAN, W. I. LIDDELL and C. J. K. WILLIAMS
3. Box and Cylindrical Columns under Biaxial Bending W. F. CHEN
4. Composite Columns in Biaxial Loading K. S. VIRDI and P. J. DOWLING
5. Cold-Formed Welded Steel Tubular Members BEN KATO
6. Buckling of Single and Compound Angles J. B. KENNEDY and M. K. S. MADUGULA
7. Centrally Compressed Built-up Structures W. UHLMANN and W. RAMM
8. Battened Columns—Recent Developments D. M. PORTER
9. Ultimate Capacity of Compression Members with Intermittent Lateral Supports PIERRE DUBAS

Index

PLATED STRUCTURES: STABILITY AND STRENGTH

edited by R. Narayanan

1. Longitudinally and Transversely Reinforced Plate Girders H. R. EVANS
2. Ultimate Capacity of Plate Girders with Openings in Their Webs R. NARAYANAN
3. Patch Loading on Plate Girders T. M. ROBERTS
4. Optimum Rigidity of Stiffeners of Webs and Flanges M. ŠKALOUD
5. Ultimate Capacity of Stiffened Plates in Compression N. W. MURRAY
6. Shear Lag in Box Girders V. KŘÍSTEK
7. Compressive Strength of Biaxially Loaded Plates R. NARAYANAN and N. E. SHANMUGAM
8. The Interaction of Direct and Shear Stresses on Plate Panels J. E. HARDING

Index

BEAMS AND BEAM COLUMNS

Stability and Strength

Edited by

R. NARAYANAN

M.Sc.(Eng.), Ph.D., D.I.C., F.I.Struct.E., F.I.C.E., F.I.E.
Senior Lecturer, Department of Civil and Structural Engineering, University College, Cardiff, United Kingdom

APPLIED SCIENCE PUBLISHERS
LONDON and NEW YORK

APPLIED SCIENCE PUBLISHERS LTD
Ripple Road, Barking, Essex, England

Sole Distributor in the USA and Canada
ELSEVIER SCIENCE PUBLISHING CO., INC.
52 Vanderbilt Avenue, New York, NY 10017, USA

British Library Cataloguing in Publication Data

Beams and beam columns.
1. Girders
I. Narayanan, R.
624.1'772 TA660.B4

ISBN 0-85334-205-9

WITH 11 TABLES AND 99 ILLUSTRATIONS

Printed in Great Britain by Galliard (Printers) Ltd, Great Yarmouth

PREFACE

I have great pleasure in writing a short preface to this book on *Beams and Beam Columns*, the second of the planned set of volumes on the stability and strength of structures.

In the United States and in Europe, we are presently writing (or rewriting) the specifications concerned with the design of thin-walled beams and beam columns. There is a vast amount of (relatively invisible) fundamental research behind most of these specifications. The object of this book is to explain the current theories which interpret the stability aspects of practical beams and beam columns and to provide the theoretical background to appreciate the specifications. As the book is addressed to structural designers and postgraduate students, a fundamental knowledge of structural mechanics is taken for granted; nevertheless, sufficient introductory material is included in each chapter to make potentially difficult subjects easily readable.

This volume contains eight chapters written by well-known experts, who have made significant contributions in the relevant subject areas. The first three chapters provide an up-to-date state-of-the-art report on the analysis and design of beams. The fourth chapter is concerned with beams with openings in webs and the necessary reinforcement. The fifth and sixth chapters discuss some analytical techniques of solving some special problems. The seventh chapter describes the behaviour of beam columns and discusses the current practices in the design of these structures. The last chapter reviews the factors affecting the selection of safety factors used in design codes. Thus the book covers a wide range of topics of relevance to the designer and the student of the subject.

I am grateful to all the contributors for the willing co-operation they have extended in producing this volume in a short time. It is hoped that the book will provide a truly international exchange of ideas and prove to be stimulating to the researcher and the engineer alike.

R. NARAYANAN

CONTENTS

Preface v

List of Contributors ix

1. Elastic Lateral Buckling of Beams 1
   D. A. NETHERCOT

2. Inelastic Lateral Buckling of Beams 35
   N. S. TRAHAIR

3. Design of Laterally Unsupported Beams 71
   D. A. NETHERCOT and N. S. TRAHAIR

4. Design of I-Beams with Web Perforations 95
   R. G. REDWOOD

5. Instability, Geometric Non-Linearity and Collapse of Thin-Walled Beams 135
   T. M. ROBERTS

6. Diaphragm-Braced Thin-Walled Channel and Z-Section Beams 161
   T. PEKÖZ

7. Design of Beams and Beam Columns 185
G. C. LEE and N. T. TSENG

8. Trends in Safety Factor Optimisation 207
N. C. LIND and M. K. RAVINDRA

Index 237

LIST OF CONTRIBUTORS

G. C. LEE

Professor and Dean, Faculty of Engineering and Applied Sciences, State University of New York at Buffalo, Buffalo, New York 14260, USA

N. C. LIND

Professor, Department of Civil Engineering, University of Waterloo, Waterloo, Ontario, Canada N2L 3G1

D. A. NETHERCOT

Senior Lecturer, Department of Civil and Structural Engineering, University of Sheffield, Mappin Street, Sheffield S1 3JD, UK

T. PEKÖZ

Associate Professor, Department of Structural Engineering, Cornell University, Ithaca, New York 14853, USA

M. K. RAVINDRA

Project Manager, Structural Mechanics Associates, 5160 Birch Street, Newport Beach, California 92660, USA

R. G. REDWOOD

Professor, Department of Civil Engineering and Applied Mechanics, McGill University, 817 Sherbrooke Street West, Montreal, PQ, Canada H3A 2K6

T. M. Roberts

Lecturer, Department of Civil and Structural Engineering, University College, Newport Road, Cardiff CF2 1TA, UK

N. S. Trahair

Professor of Civil Engineering, School of Civil and Mining Engineering, University of Sydney, Sydney, New South Wales 2006, Australia

N. T. Tseng

Research Assistant Professor, Faculty of Engineering and Applied Sciences, State University of New York at Buffalo, Buffalo, New York 14260, USA

Chapter 1

ELASTIC LATERAL BUCKLING OF BEAMS

D. A. NETHERCOT

Department of Civil and Structural Engineering,
University of Sheffield, UK

SUMMARY

The mechanism by which slender beams buckle by a combination of lateral bending and twisting is described and the basic theoretical solution presented. The ways in which this may be modified to allow for variations in the type of loading and conditions of lateral support are discussed. Extensions to the basic theory to account for the influence of pre-buckling deflections, post-buckling strength, monosymmetric cross-sections or variable section properties are summarised. Finally, the elastic behaviour of geometrically imperfect beams is discussed as a prelude to the treatment of design procedures in Chapter 3.

NOTATION

a	Distance of point of cross-section from shear centre
$\bar{a}$	Distance of transverse load from shear centre axis
b	Distance of transverse load from support
f	Stress
h	Distance between flange centroids
k	Effective length factor
k_y, k_w	Effective length factors for minor axis, warping restraint
l	Effective length
m	Equivalent uniform moment factor
n	Constant

r	Reduction factor
t	Web thickness
u	Deflection of shear centre in x-direction
u_0	Initial lateral deflection
w	Uniformly distributed load
x, y	Major and minor principal axes of section
$\bar{y}$	Distance from flange centroid to section centroid
y_0	Coordinate of shear centre
z	Longitudinal axis through centroid
A_1, A_2	Constants
B	Flange width
B_B, B_T	Width of bottom and top flanges
D	Depth of section
D_L	Depth of lip
E	Young's modulus of elasticity
F_b	Bending stress about major axis $= M/Z_x$
F_{max}	Maximum longitudinal stress
F_Y	Yield stress
F_{0b}	Elastic buckling stress $= M_E/Z_x$
G	Shear modulus of elasticity
I_w	Warping section constant
I_x, I_y	Second moments of area about the x, y axes
I_{yc}, I_{yt}	Second moments of area of compression, tension flanges about the y-axis
J	Torsion section constant
K	Beam parameter (eqn (1.7))
$\bar{K}$	Torque component for monosymmetric beams (eqn (1.23))
L	Span or segment length
M	Moment
M_B, M_T	Restraining moment on bottom or top flange
M_c	Buckling moment
M_E	Elastic buckling moment
M_{Er}	Reduced elastic buckling moment (eqn (1.19))
M_x	Major-axis bending moment
M_z	Restraining moment about z-axis
M_0	Elastic buckling moment for a simply supported beam in uniform bending (eqn (1.5))
M_{0m}	Value of M_0 for monosymmetric beam (eqn (1.28))
P_y	Elastic critical load for buckling as a strut about the minor axis

R, R_2, R_3, R_4	End restraint parameters
T	Flange thickness
U	Strain energy
W	Transverse concentrated load
Z_x, Z_y	Section modulus about the x, y axes
α_c, α_R	Stiffness of critical, restraining segments
α_z	Torsional stiffness of end support
β	Ratio of end moments
β_x	Monosymmetry parameter (eqn (1.24))
γ	Beam cross-sectional parameter (eqn (1.29))
δ_0	Maximum initial lateral deflection
λ_c, λ_R, λ_F	Buckling load factors (eqn (1.20))
ξ, η, ζ	Deformed axis system (Fig. 1.3)
θ_0	Maximum initial twist
ϕ	Angle of twist
ϕ_0	Initial twist
ρ	Ratio of second moment of area of compression flange to second moment of area of cross-section

1.1 INTRODUCTION

The main design requirement for a beam is normally bending strength about its major axis. In the case of steel members this leads naturally to the use of I-sections, with most of the material concentrated in the flanges away from the neutral axis. Considerations of possible local instability mean that these flanges must not be too wide (a limitation is usually placed on their width to thickness ratio), leading to the narrow flange type of rolled section shown in Figs. 1.1(a) and 1.1(b). For large beams, where strengths in excess of those that can be provided by the normal range of rolled sections are required, e.g. bridge beams, crane girders, etc., bigger versions of this general shape must be specially fabricated—usually by welding together three plates as shown in Fig. 1.1(c).

When a narrow flange I-section beam is carefully tested under laboratory conditions its initial response will be to deflect vertically. Because of its high in-plane bending stiffness, such deflections will not normally be large, especially whilst the material in the beam remains elastic. However, at a certain critical value of the applied load the beam will fail suddenly by deflecting sideways and twisting as illustrated in Fig. 1.2. Such behaviour,

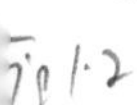

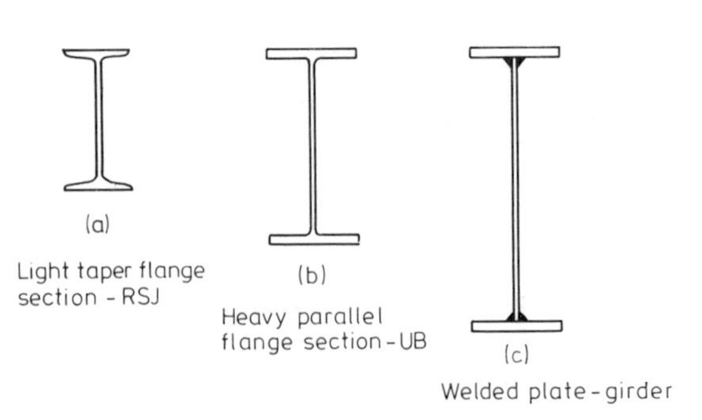

FIG. 1.1. I-sections used as beams.

which is termed lateral or lateral–torsional buckling, represents an upper limit on the beam's effective load-carrying capacity. The phenomenon has much in common with the Euler buckling of a strut; in both cases a member that has been loaded in a stiff plane (axially for the strut, in major-axis bending for the beam) buckles in a weaker plane (flexurally for the strut, in combined torsion and minor-axis flexure for the beam).

The earliest theoretical analyses of lateral stability were produced independently by Prandtl (1899) and Michell (1899). Both authors restricted themselves to a beam of narrow rectangular cross-section, simply supported at either end, loaded by uniform moment. Some years later Timoshenko (1905) extended these solutions to I-sections, by including the effects of warping on the torsional aspects of the problem. Since then many authors have developed the topic, and the influence of the conditions of loading and lateral restraint, monosymmetry (which further complicates torsional behaviour), non-uniform sections and inelastic material behaviour have all been studied. Previous reviews of the subject have been published by Lee (1960), Nethercot and Rockey (1971), Galambos (1977) and Trahair (1977).

This chapter should be regarded as the first of a series of three dealing with the lateral torsional buckling of steel beams. It is limited to elastic buckling; inelastic buckling is covered in Chapter 2 and design procedures for laterally unbraced beams are reviewed in Chapter 3. Although the material presented in this chapter is almost entirely related to the theoretical treatment of the problem—elastic buckling tests are reviewed in Chapter 3—a full mathematical coverage is not provided. Rather the main aspects of the theory are discussed in the context of the assumptions made and the results obtained. Readers wishing to acquaint themselves with the full derivations and the mathematical processes employed in the solutions

FIG. 1.2. Lateral buckling of a cantilever beam.

should refer either to basic textbooks (Timoshenko and Gere, 1961; Chen and Atsuta, 1977; Kirby and Nethercot, 1979) or to the original research papers as appropriate.

1.2 BUCKLING OF UNIFORM EQUAL-FLANGED I-BEAMS

1.2.1 Buckling of Simply Supported Beams in Uniform Bending

(a) Basic Theory

Figure 1.3 illustrates the generally accepted basic lateral buckling problem. The beam, which is of doubly symmetrical cross-section, is loaded by equal

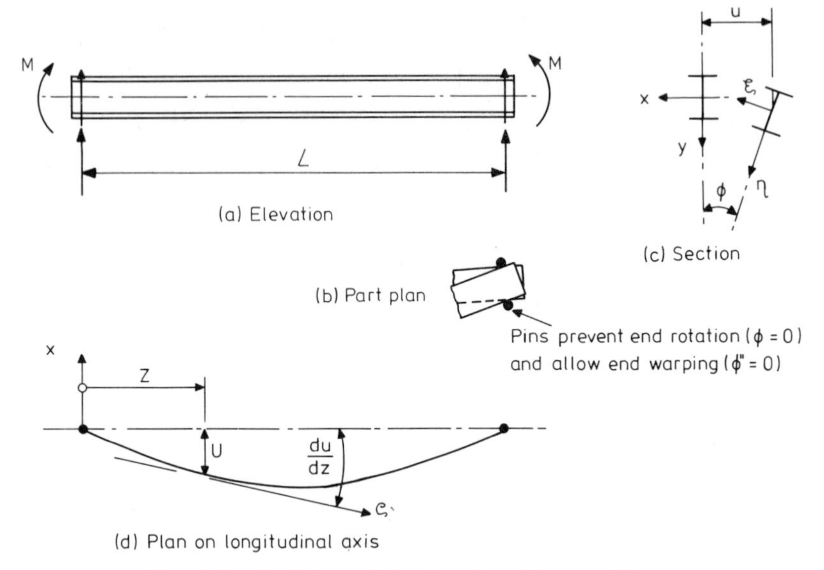

FIG. 1.3. Basic lateral–torsional buckling problem.

and opposite end moments M, so as to cause bending about its major principal axis. At both ends it is supported in such a way that lateral deflection and twist are prevented but no resistance is offered to either lateral bending or warping (axial movement accompanying torsion which corresponds to bending of the flanges in opposite senses about an axis through the web). Placing the beam in its slightly buckled state, i.e. allowing the lateral u and torsional ϕ deformations to occur and resolving M into its components about the deformed axis system ξ, η, ζ, enables the

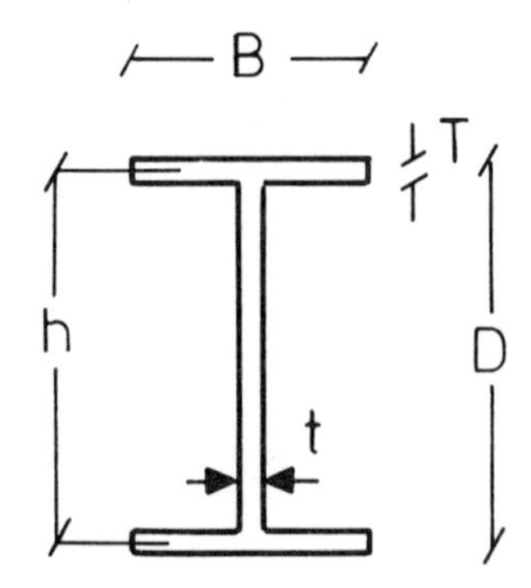

FIG. 1.4. Approximation of I-section used to calculate cross-sectional properties.

disturbing moments to be equated to the beam's internal resistance to give (Timoshenko and Gere, 1961; Chen and Atsuta, 1977; Kirby and Nethercot, 1979):

$$EI_y \frac{\mathrm{d}^2 u}{\mathrm{d}z^2} = M\phi \tag{1.1}$$

$$GJ \frac{\mathrm{d}\phi}{\mathrm{d}z} - EI_w \frac{\mathrm{d}^3\phi}{\mathrm{d}z^3} = \frac{\mathrm{d}u}{\mathrm{d}z} \tag{1.2}$$

Solving these two equations subject to the boundary conditions

$$u = \frac{\mathrm{d}^2 u}{\mathrm{d}z^2} = 0 \qquad \text{at } z = 0, L \tag{1.3}$$

$$\phi = \frac{\mathrm{d}^2 \phi}{\mathrm{d}z^2} = 0 \qquad \text{at } z = 0, L \tag{1.4}$$

gives the expression for the least value of M necessary to cause elastic instability, M_0, as

$$M_0 = \frac{\pi}{L}\sqrt{(EI_y GJ)}\sqrt{\left(1 + \frac{\pi^2 EI_w}{L^2 GJ}\right)} \tag{1.5}$$

The solution also provides information on the buckled shape—it is sinusoidal in both u and ϕ—but because of the way the problem has been set up it cannot give any indication of the magnitude of the deformations involved.

(b) Physical Significance of Solution

Inspection of the terms in eqn (1.5) shows that a beam's elastic critical moment depends upon its minor-axis flexural rigidity EI_y, its torsional rigidity GJ and its warping stiffness EI_w. Approximating a symmetrical I-section as three rectangles as shown in Fig. 1.4 enables these to be calculated from

$$\begin{aligned} I_y &= TB^3/6 \\ J &= \tfrac{1}{3}(2BT^3 + (D-T)t^3) \\ I_w &= I_y(D-T)^2/4 \end{aligned} \tag{1.6}$$

More exact formulae for J, which allow for the radii and fillets present in rolled sections, have been provided by Johnston and El Darwish (1965), whilst tabulated values for standard sections are available (Terrington, 1968, 1970; AISC, 1969). Procedures for the determination of J and I_w for more complex shapes have also been published (Timoshenko and Gere, 1961; Zbirohowski-Koscia, 1967; BSI, 1969).

Length is also an important influence on lateral stability; eqn (1.5) contains both a direct relationship between M_0 and $1/L$ as well as a further connection via the terms under the second square-root sign. Putting

$$\sqrt{\left(\frac{\pi^2 EI_w}{L^2 GJ}\right)} = K \tag{1.7}$$

enables eqn (1.5) to be rewritten as

$$M_0 = \frac{\pi}{L}\sqrt{(EI_y GJ)}\sqrt{(1 + K^2)} \tag{1.8}$$

Thus K is a measure of the relative importance of the two mechanisms for resisting the torsional component of lateral buckling, i.e. the two terms on the left-hand side of eqn (1.2); it is also dependent upon L. Values of K for practical beams will normally lie towards the centre of the range $0{\cdot}1 < K < 3{\cdot}0$. Generally, low values are associated with long beams of fairly compact cross-section, e.g. column-type shapes, and high values correspond to short beams of slender cross-section, e.g. deep plate-girders.

1.2.2 Buckling of Simply Supported Beams in Non-uniform Bending

(a) General

The solution of eqns (1.1) and (1.2) in the previous section was facilitated by the limitation to loading comprising uniform moment M. For other forms of loading, e.g. unequal end moments, a uniformly distributed transverse load, etc., these differential equations will no longer contain constant coefficients (the value of M at any point will now be a function of z) with the result that numerical methods of solution will normally be required. Techniques that have successfully been applied include the methods of finite differences (Vinnakota, 1977), finite elements (Barsoum and Gallagher, 1970; Nethercot, 1972*a*) and finite integrals (Vacharajittiphan and Trahair, 1974; Brown and Trahair, 1975; Trahair, 1977). As an alternative to the formation and solution of the governing equations, the problem may be studied using the energy method, if a close approximation of the buckled shape can be devised.

In this case the work done by the loads as the beam buckles is equated to the strain energy stored during buckling, which is given by (Timoshenko and Gere, 1961; Barsoum and Gallagher, 1970):

$$U = \frac{1}{2}\int_0^L EI_y\left(\frac{d^2u}{dz^2}\right)dz + \frac{1}{2}\int_0^L GJ\left(\frac{d\phi}{dz}\right)^2 dz + \frac{1}{2}\int_0^L EI_w\left(\frac{d^2\phi}{dz^2}\right)^2 dz \tag{1.9}$$

For the particular case of uniform moment loading, the beam buckles in a half-sinewave for both lateral deflection u and twist ϕ, and most of the strain energy associated with flexure (EI_y) and warping (EI_w) is therefore stored in the central portion of the span where d^2u/dz^2 and $d^2\phi/dz^2$ are high.

(b) Equivalent Uniform Moment Factor
Provided M_E is regarded as the critical value of the maximum moment in the beam, it is possible to express the results for other forms of loading in a slightly modified form of eqn (1.5) as

$$M_E = m\frac{\pi}{L}\sqrt{(EI_yGJ)}\sqrt{\left(1+\frac{\pi^2 EI_w}{L^2GJ}\right)} = mM_0 \tag{1.10}$$

in which m = equivalent uniform moment factor. Equation (1.10) provides a clear indication of the relative severity (in terms of its effect on lateral stability) of any given loading arrangement through the value of m. Since uniform moment is the 'worst case', such values will be greater than unity.

In many instances the value of m has been found (Nethercot and Rockey, 1971, 1973) to be virtually independent of the proportions of the beam, and Table 1.1 lists these cases.

(c) Beams Subject to Moment Gradient
Of particular importance is the case of the beam subject to end moments M and βM where $-1{\cdot}0 \leq \beta \leq 1{\cdot}0$, since loads are frequently transmitted to main beams by secondary members which prevent lateral movement of the loaded point as shown in Fig. 1.5. For this case m does show some small

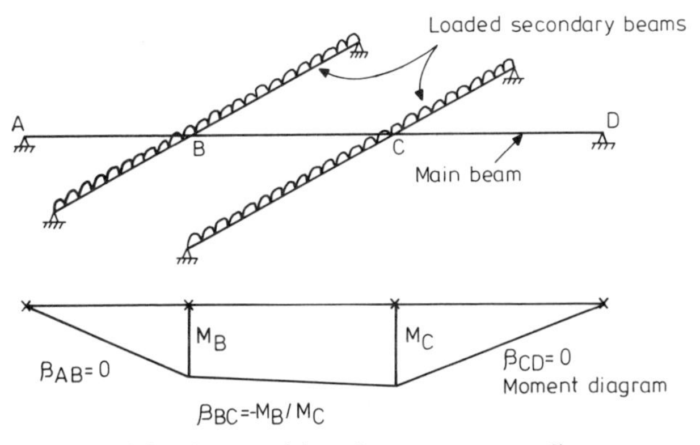

FIG. 1.5. Beam subjected to moment gradient.

TABLE 1.1
EQUIVALENT UNIFORM MOMENT FACTORS m

Beam and Loads	Bending Moment	M_{max}	m
 M M		M	1·00
 M		M	1·75
 M M		M	2·56
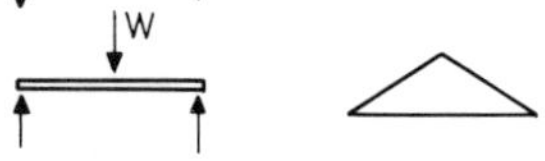 W		$\frac{WL}{4}$	1·35
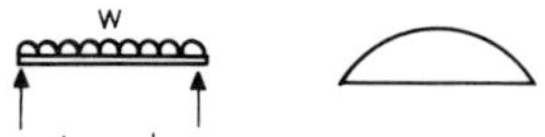 w		$\frac{wL^2}{8}$	1·13
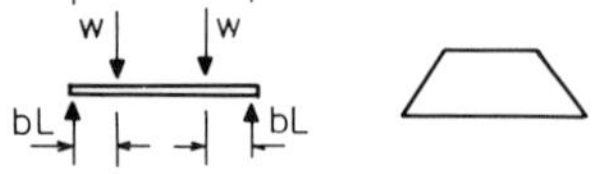 w w bL bL		wbL	$1+b^2$
 w bL		$wb(1-b)L$	$1{\cdot}35+1{\cdot}95(0{\cdot}5-b)^2$

dependence on K, particularly as the moment gradient increases, i.e. $\beta \rightarrow 1{\cdot}0$. However, for design purposes this may normally be ignored so that m can be taken as a function of β only. Of the various formulae proposed (Horne, 1954; Salvadori, 1955, 1956) the most commonly accepted are

$$m = 1{\cdot}75 + 1{\cdot}05\beta + 0{\cdot}3\beta^2 \leq 2{\cdot}56 \tag{1.11}$$

$$\frac{1}{m} = 0{\cdot}6 - 0{\cdot}4\beta \geq 0{\cdot}4 \tag{1.12}$$

Both are compared with the accurate solution in Fig. 1.6. The reason for the reciprocal format of eqn (1.12) is that design recommendations are sometimes framed such that the actual moments are scaled down by $1/m$ rather than the available strength being factored up by m.

(*d*) *Beams Carrying Transverse Loads*

In the case of beams loaded by transverse loads the preceding discussion has assumed that such loads act at the level of the beam's centroid, which

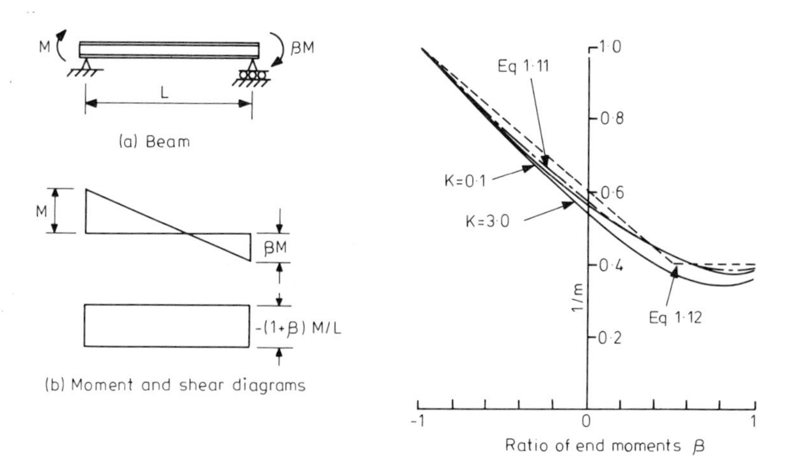

FIG. 1.6. *m*-factors for moment gradient loading.

for the doubly symmetrical type of section under consideration coincides with the shear centre. If such loads are free to move sideways with the beam as it buckles, then altering the level at which they act can significantly influence their effect on the stability of the beam (Nethercot and Rockey, 1971; Nethercot, 1972*b*,*c*). As shown in Fig. 1.7, placing the load at a distance $\bar{a}$ above the shear centre causes it to exert an additional torque $W\bar{a}\phi$ which decreases the resistance of the beam to buckling. Conversely, loads applied below the level of the shear centre are less dangerous owing to the stabilising influence of this additional torque. The magnitude of this load height effect depends upon both the type of loading, e.g. pair of point loads, uniformly distributed load, etc., as well as the value of the beam torsional parameter K. Figure 1.7 illustrates this for the case of a central point load, whilst Table 1.2 provides expressions (Nethercot and Rockey, 1971; Nethercot, 1972*b*,*c*) for determining good estimates of the *m*-factors for several other cases.

1.2.3 Buckling of End Restrained Beams

(a) General

Figure 1.3 illustrated the type of end support conditons assumed in the basic lateral–torsional buckling problem. However, beams forming parts of structures will often be provided with different forms of end restraint. When such restraints act about the major axis, they influence lateral

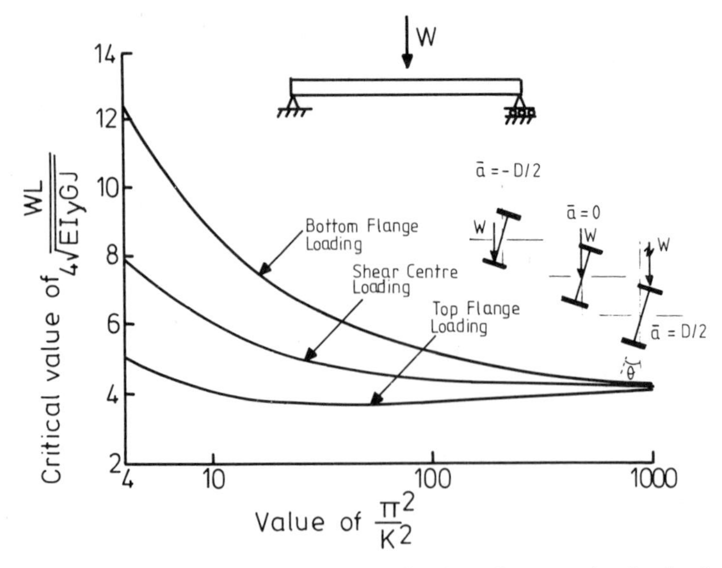

FIG. 1.7. Effect of the level of application of a central point load.

stability through their effect on the pattern of moments present in the beam. Restraints against minor-axis bending and against torsion, on the other hand, exert a more direct influence because they affect the buckled shape. In the case of the basic conditions the former are assumed to be absent, whilst the latter are assumed to be fully effective.

TABLE 1.2
VALUES OF m FOR FLANGE LOADING

Load at top flange $m = m_s / n$

Load at bottom flange $m = m_s \times n$

Where m_s = value of m for that type of loading applied at the level of the centroid

Type of loading	Value of $\bar{n}$
W (uniformly distributed)	$1 - 1{\cdot}54K^2 + 0{\cdot}535K$
W (central point load)	$1 - 0{\cdot}180K^2 + 0{\cdot}649K$
bL W, W bL (two point loads)	$1 - 0{\cdot}465K^2 + 1{\cdot}636K$

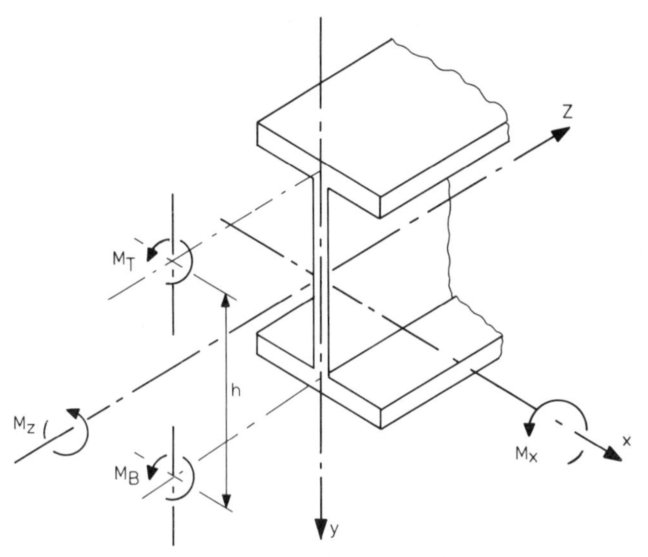

FIG. 1.8. End restraining actions for beams.

The four possible forms of end restraint illustrated in Fig. 1.8 are:

(i) restraint against major-axis bending M_x;
(ii) restraint against rotation of the bottom flange M_B;
(iii) restraint against rotation of the top flange M_T;
(iv) restraint against twist M_z.

Types (ii) and (iii) cover both minor-axis bending and warping respectively through the restraint parameters:

$$R_2 = \frac{M_B + M_T}{[M_B + M_T]_{fe}} \tag{1.13}$$

$$R_4 = \frac{(M_T - M_B)/2}{[(M_T - M_B)/2]_{fe}} \tag{1.14}$$

where the subscript fe denotes the value of that quantity corresponding to full fixity, i.e. R_2 and R_4 may vary between zero when no restraint is present and 1 when minor-axis rotation and/or warping is prevented.

In the same way that the buckling of a restrained column may be related to the buckling of a similar pin-ended column through the idea of 'effective length', so the critical load for a beam with end restraint may be written as

$$M_E = \frac{\pi}{l}\sqrt{(EI_y GJ)}\sqrt{\left(1 + \frac{\pi^2 EI_w}{l^2 GJ}\right)} \tag{1.15}$$

in which $l = kL$ is the effective length. For the particular case of symmetrical restraints arranged such that $R_2 = R_4 = R$ with loading corresponding to uniform moment, the relationship between the effective length factor k and the end restraint parameter R is closely approximated by

$$k = 1 - 0{\cdot}5R \tag{1.16}$$

provided R is defined by

$$\frac{\text{Flange end moment}}{\text{Flange end rotation}} = -\frac{EI_y}{L}\left(\frac{R}{1-R}\right) \tag{1.17}$$

Because l appears twice in eqn (1.15) the increase in stability due to full rotational restraint in plane, i.e. $k = 0{\cdot}5$ corresponding to $R = 1$ in eqns (1.13), (1.14) and (1.16), will vary from twice when K is small to nearly four times when K is large.

For cases in which $R_2 \neq R_4$ recourse should be made to the tabulated values of elastic critical loads that are available (Austin *et al.*, 1955; Trahair, 1965, 1968*a*, 1969). When lateral bending and/or warping is prevented at one or both ends, Nethercot (1972*b*,*c*) and Nethercot and Rockey (1973) provide solutions in the form of *m*-factors for use with eqn (1.10) for each of the load cases shown in Table 1.1.

(*b*) *Incomplete Torsional Restraint*

In the basic problem of Fig. 1.3, end twisting is assumed to be prevented. However, situations may arise for which the torsional resistance of the end supports is such that this condition cannot be met, i.e. α_z in the following expression is less than infinite:

$$\alpha_z = \frac{GJ}{L}\frac{1}{R_3} \tag{1.18}$$

in which α_z = torsional stiffness of end supports. Provided R_3 is more than about 10 or 15, a figure that is not difficult to achieve in practice, then the corresponding reduction in buckling moment M_{Er} will not be more than a few per cent. The relationship between M_{Er} and R_3 may be approximated (Trahair, 1965) within the range $0{\cdot}9 \leq M_{Er}/M_E \leq 1{\cdot}0$ by

$$M_{Er}/M_E = 1 - R_3^{0{\cdot}95}[A_1 + A_2(K)^{0{\cdot}95}] \tag{1.19}$$

in which A_1 and A_2 are constants which both take the value 1·1 for a beam with a central point load.

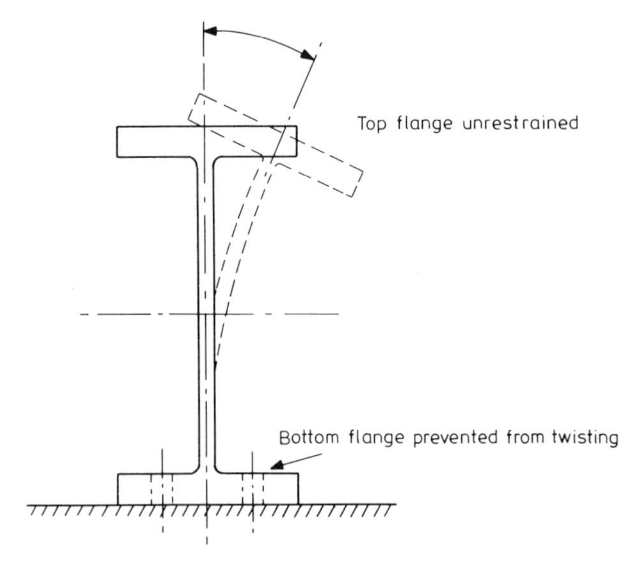

FIG. 1.9. Incomplete end torsional restraint.

Incomplete end torsional restraint can also arise for the arrangement shown in Fig. 1.9, in which the top flange of the beam is left laterally unrestrained at the support. Lateral buckling may now be accompanied by distortion of the cross-section, particularly for deep beams with thin webs, resulting in generally lower buckling loads. A theoretical method for studying this problem has been published by Bartels and Bos (1973) who also conducted experiments which dramatically illustrated the potential dangers of this type of support. As a general guide, for long shallow beams the problem is essentially one of lateral–torsional buckling under an inferior end restraint, whereas for short deep beams the situation corresponds rather more to local buckling of a portion of the web under the action of a concentrated vertical force, e.g. the end reaction.

(c) Cantilever Beams

One important class of problem for which both the loading and the support conditions differ from those illustrated in Fig. 1.3 is the cantilever. Solutions are available for a variety of loading conditions, including transverse loads applied at the level of either flange (Anderson and Trahair, 1972; Nethercot, 1973*a*). Despite the fact that the upper flange is now the tension flange, it is still the flange that deforms more in the buckled shape[2] (see Fig. 1.2) and lateral stability is still reduced as the load level is raised relative to the beam's shear centre, as illustrated in Fig. 1.10.

The buckling of cantilevers provided with various forms of restraint at the tip, including cases where the cantilever is formed by an overhanging portion of a continuous beam and different lateral restraint conditions are therefore possible at the root, has also been studied (Nethercot, 1973*a*).

Although Nethercot (1973*a*) provides expressions for determining *m*-factors for use with eqn (1.9) for each of these cases (L being taken as the full span), the alternative of using a modified effective length l is the more usual simplification. The values presented in Table 1.3 for use with eqn (1.15) have been simplified to the extent that where k varies with K, only a minimum value within the practical range of K values has been given.

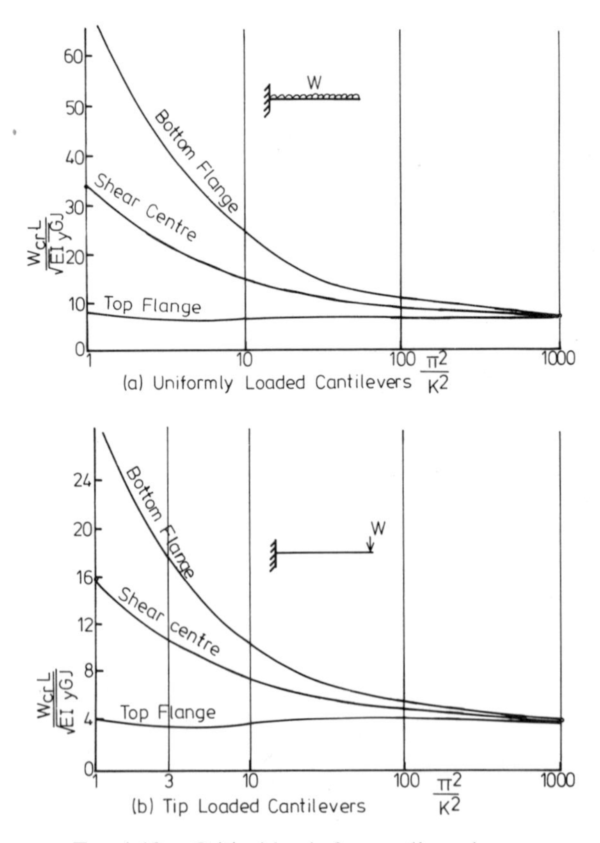

FIG. 1.10. Critical loads for cantilever beams.

TABLE 1.3
EFFECTIVE LENGTHS FOR CANTILEVERS

Restraint conditions		Effective length	
At root	At tip	Top flange loading	All other cases
		1·4L	0·8L
	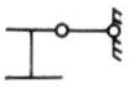	1·4L	0·7L
	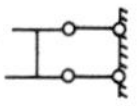	0·6L	0·6L
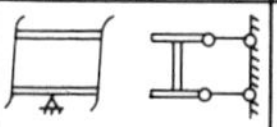		2·5L	1·0L
	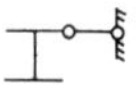	2·5L	0·9L
	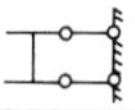	1·5L	0·8L
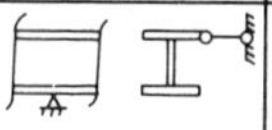		7·5L	3·0L
	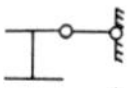	7·5L	2·7L
	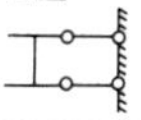	4·5L	2·4L

Both end load and uniformly distributed loading have been considered; this explains why different values are provided for centroidal and top flange loading when twisting of the tip is prevented.

1.2.4 Buckling of Laterally Continuous Beams

(a) General

The buckling of beams which are continuous in the lateral plane will normally involve interaction between adjacent segments. (In this context it is unimportant whether such beams are single-span beams divided into a number of segments by a series of rigid lateral restraints, as shown in Fig. 1.11(a), or whether they are also continuous in the vertical plane, as shown in Fig. 1.11(b).) If both the lengths of the individual segments and the patterns of moments within them are similar, then this interaction may also be slight. However, situations involving considerable differences between adjacent segments may well result in the less critical segments being capable of providing significant restraint to the most critical segment. A full explanation of the mechanism of this 'interaction buckling' has been provided by Trahair (1968*b*).

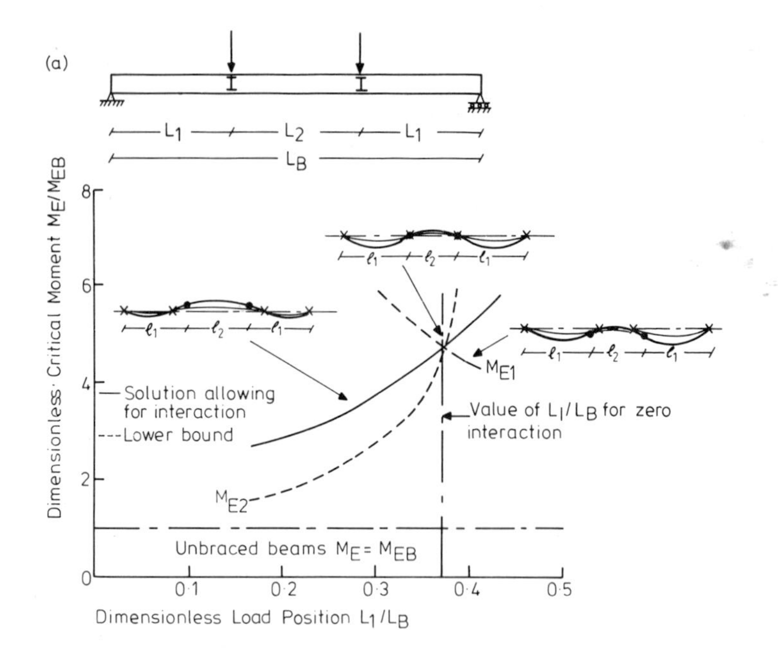

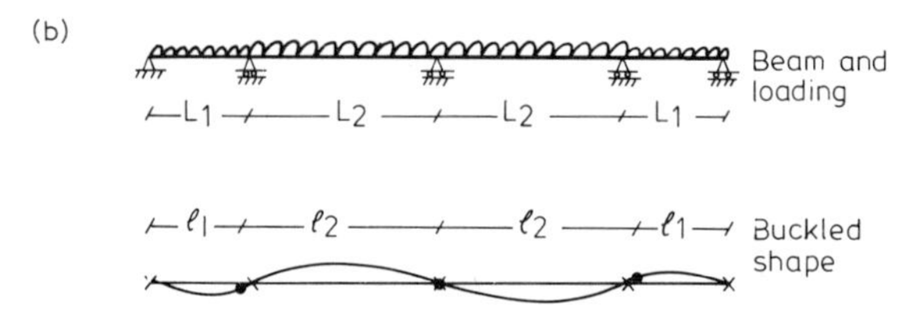

FIG. 1.11. Buckling of continuous beams.

(b) Lower Bound Approach

The simplest approach to the analysis of this problem is that first proposed by Salvadori (1951), which ignores the effects of continuity between adjacent segments. It therefore provides a lower bound estimate of the critical load for the whole beam determined as the value of the load parameter at which the most critical segment considered as an isolated laterally simply supported beam would become unstable. In cases where the location of the most critical segment is not obvious, each of the possible segments should be considered; the correct one will be that which possesses the lowest individual critical load. In evaluating critical loads for individual segments, use should be made of the m-factors of Tables 1.1 and 1.2. In simple physical terms, Salvadori's method assumes a buckled shape in which the points of inflection are located at the points of lateral restraint.

(c) Allowing for Continuity

The lower bound estimate may be improved upon by allowing for the restraining influence of the more stable segments on either side of the critical segment. Based on the observation that the relationship between restraint stiffness and effective length factor k for a symmetrically restrained beam (discussed in Section 1.2.3) is identical to that for a similarly restrained strut, Nethercot and Trahair (1976, 1977) have suggested the use of a column effective length chart, of the type shown as Fig. 1.12, as a means of estimating the effective lengths of unsymmetrically restrained beam segments. Reductions in the effectiveness of such restraint due to the presence of the destabilising major-axis moments in these adjacent segments may be approximated by using an amplification factor to give

$$\alpha_R = n\left(\frac{EI_y}{L}\right)_R\left(1 - \frac{\lambda_c}{\lambda_R}\right) \tag{1.20}$$

in which n = a constant whose value depends on the end conditions as explained in Fig. 1.13, λ_c = buckling load factor for the critical segment, obtained using Salvadori's method, and λ_R = buckling load factor for the

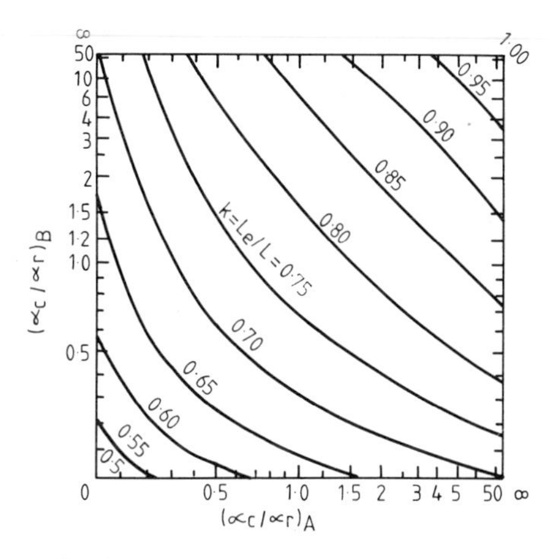

FIG. 1.12. Effective length chart for restrained beam segments.

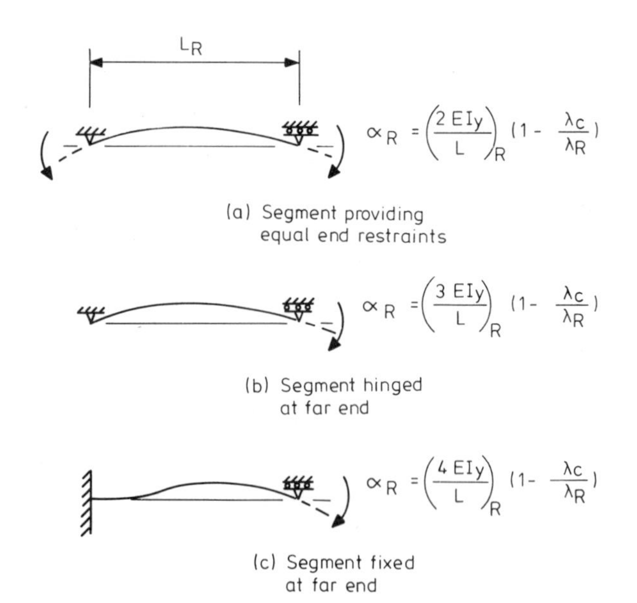

FIG. 1.13. Stiffness of restraining segments for use with eqn (1.20).

restraining segment, obtained using Salvadori's method. For the critical segment α_c should be taken as

$$\alpha_c = \left(\frac{2EI_y}{L}\right)_c \tag{1.21}$$

Use of this method involves little additional calculation beyond that needed for the traditional lower bound approach since the effective length determination is made only for the critical segment.

Several improvements to the above technique have been suggested by Dux and Kitipornchai (1978). These include: use of a series of effective length charts which permit the dependence of k on both moment gradient β and beam geometry K to be recognised, a parabolic reduction in the restraining effect of adjacent segments with increasing major-axis moment, use of a value λ_c and λ_R when determining α_R, and cycling of the effective length determination until the assumed λ_F and the resulting beam critical load parameter coincide. Although all of these procedures can be justified on the grounds that they approximate better the actual situation, their introduction does complicate the process. However, accuracy is improved, particularly for beams containing segments with high moment gradients ($\beta \rightarrow 1{\cdot}0$).

Further improvements for the case of beams loaded so as to produce moment patterns that are antisymmetrical about braced points have been suggested by Cuk and Trahair (1981). The effect of the warping restraint present at such braced points (which tends to be neglected by both the other methods) is determined by separating the effective length factor k into k_y, which covers minor-axis bending restraint, and k_w, covering warping restraint. Values of k_w as a direct function of the shape of the moment diagram in the segments on either side of the brace were obtained by curve-fitting. Some idea of the accuracy of each of these methods may be obtained from the example problem shown in Fig. 1.14.

1.2.5 Refinements to the Basic Theory

(a) Effect of In-Plane Deflection

In deriving eqns (1.1) and (1.2) and hence eqn (1.5), any effect of major-axis deflection was neglected. When this is included (Trahair and Woolcock, 1973), the expression for the critical moment for the basic problem of a simply supported beam subject to equal end moments (see Fig. 1.3) becomes

$$M_E = \frac{\pi}{L} \frac{\sqrt{\left(EI_y GJ\left(1 + \dfrac{\pi^2 EI_w}{L^2 GJ}\right)\right)}}{\sqrt{\left(\left(1 - \dfrac{EI_y}{EI_x}\right)\left[1 - \dfrac{GJ}{EI_x}\left(1 + \dfrac{\pi^2 EI_w}{GJL^2}\right)\right]\right)}} \tag{1.22}$$

This reduces to eqn (1.5) when EI_y is large. For a typical beam-type section ($I_x/I_y \simeq 10$) and a typical column-type section ($I_x/I_y \simeq 3$), the percentage increases in M_E are approximately 5 and 23 respectively; in both cases these figures are virtually independent of beam span.

Such increases might appear to suggest that to ignore this effect would be to err unnecessarily on the conservative side. However, good reasons exist to limit its inclusion to those situations where its beneficial effect is certain to be present. The first of these is the practical consideration that any precambering of the beam to reduce its deflection will of course reduce, or even perhaps remove, the effect of in-plane deflection. A second, and more fundamental, reason is that the magnitude of the effect is heavily dependent upon the conditions of lateral restraint provided (Vacharajittiphan *et al.*, 1974), although it is not very sensitive to the pattern of moments. As indicated in Table 1.4, which gives average percentage increases in M_E for a variety of arrangements, figures much smaller than those given previously are possible. Indeed for certain cases

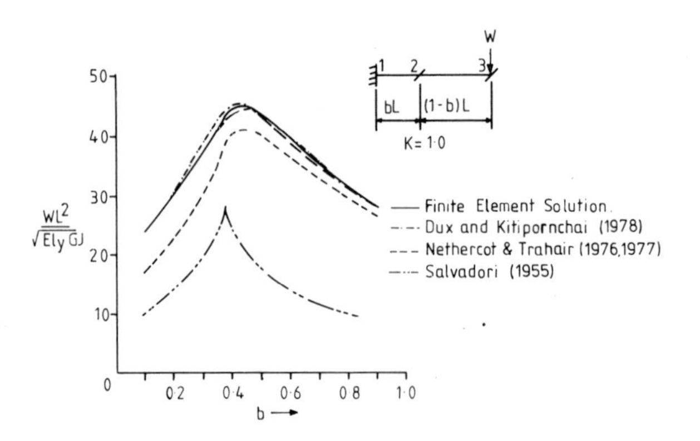

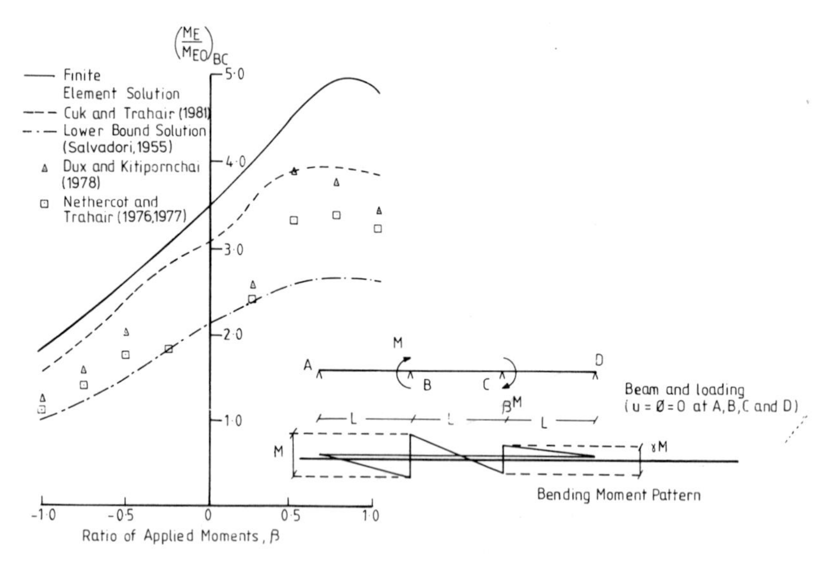

FIG. 1.14. Comparison of approximate methods of determining the buckling loads of continuous beams.

TABLE 1.4
PERCENTAGE INCREASES IN CRITICAL LOADS DUE TO IN-PLANE DEFLECTION

Moment pattern	End restraint condition (plan view)	Percentage increase in M_E due to in-plane deflection	
		Beam-type $I_x/I_y \approx 10$	Column-type $I_x/I_y \approx 3$
		5	23
		5	23
		5	23
		5	24
		-4	-5
		0	4
		4	16

M_E actually decreases. In the case of transverse loads free to move with the beam on buckling, the level of application also becomes significant (Vacharajittiphan *et al.*, 1974); smaller increases are obtained for top flange loading, particularly for short beams of column-type proportions.

(*b*) *Post-buckling Analysis*

In common with the basic Euler theory for struts, eqns (1.1) and (1.2) are based on the use of the so-called 'small deflection' theory. As indicated previously, this provides no information on the magnitude of the buckling deformations. Woolcock and Trahair (1974) have reformulated the problem so that the full load-deformation behaviour for loads in excess of the theoretical elastic critical load may be followed. Results for both simply supported beams and cantilevers show that lateral deflection and twist increase rapidly for loads only slightly beyond the initial buckling load. Moreover, the maximum longitudinal stresses in the beams also increase sharply so that for practical purposes no additonal load-carrying capacity is available in the post-buckling region, even for extremely slender beams.

This situation changes dramatically if, through the presence of in-plane restraint, redistribution of the major-axis moment pattern becomes possible. Early work by Masur and Milbradt (1957), subsequently confirmed and extended by Woolcock and Trahair (1976), has shown that slender beams of narrow rectangular cross-section may support loads of up to three times their initial buckling loads. However, the beam geometries are outside the practical range, and it would appear that the corresponding increases for beams of the proportions common in civil engineering structures may be much smaller.

1.3 BUCKLING OF MONOSYMMETRIC BEAMS

1.3.1 Extensions to the Basic Theory for Beams in Uniform Bending

The theory developed in Section 1.2 for doubly symmetrical sections is also applicable to sections symmetrical about the minor axis only, e.g. channels, provided the applied loading acts through the shear centre (which is still situated on the axis of symmetry although not at the same point as the centroid). For sections symmetrical about the major axis only, e.g. unequal flanged I's, however, the non-coincidence of the shear centre and the centroid complicates the torsional behaviour of the beam.

Assuming the loading to correspond to uniform moment, the resulting axial stresses produce a torque about the shear centre given by

$$\bar{K}\frac{\mathrm{d}\phi}{\mathrm{d}z} = \frac{\mathrm{d}\phi}{\mathrm{d}z}\int^{A} fa^2\,\mathrm{d}A \tag{1.23}$$

in which f = longitudinal stress at a point distance a from the shear centre.

Defining the position of the shear centre by y_0 as shown in Fig. 1.15 and noting that $f = M/I_x$ enables eqn (1.23) to be rewritten as

$$\bar{K}\frac{\mathrm{d}\phi}{\mathrm{d}z} = \frac{\mathrm{d}\phi}{\mathrm{d}z} M\left(\frac{\int^{A} x^2 y\,\mathrm{d}A + \int^{A} y^3\,\mathrm{d}A}{I_x} - 2y_0\right) \tag{1.24}$$

The expression within brackets is often termed β_x; it is a monosymmetry parameter and an explicit expression for its evaluation in the case of a monosymmetric I-section is provided in Fig. 1.15. Studies by Kitipornchai and Trahair (1980) have shown that the simpler alternatives of eqn (1.25)

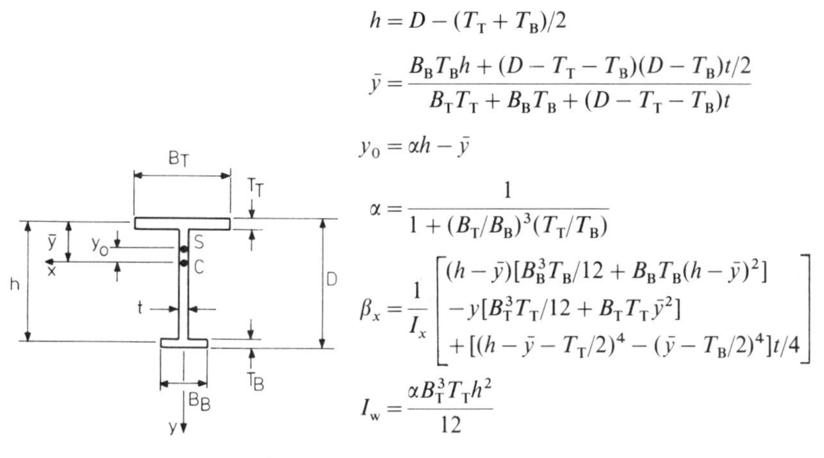

FIG. 1.15. Sectional properties of monosymmetric I-beams.

for plain I's or eqn (1.26) for I's with a lipped compression flange are sufficiently accurate for the majority of cases:

$$\beta_x = 0{\cdot}9h(2\rho - 1)[1 - (I_y/I_x)^2] \tag{1.25}$$

$$\beta_x = 0{\cdot}9h(2\rho - 1)[1 - (I_y/I_x)^2][1 + D_L/2D] \tag{1.26}$$

in which $\rho = I_{yc}/(I_{yc} + I_{yt})$ and D_L = depth of lip.

Inclusion of this effect modifies eqn (1.2) to

$$(GJ + M\beta_x)\frac{d\phi}{dz} - EI_w\frac{d^3\phi}{dz^3} = M\frac{du}{dz} \tag{1.27}$$

Thus the principal effect of monosymmetry is to cause the beam's effective torsional rigidity to be increased when the larger flange is in compression or decreased when the smaller flange is in compression (y_0 is positive and β_x becomes negative).

This situation is therefore similar to the so-called Wagner effect for columns. It results in the value of the critical moment for an unequal flanged beam being dependent upon which of the flanges is in compression through the presence of β_x in the expression for M_{0m}, viz.

$$M_E = \sqrt{\left(\frac{\pi^2 EI_y}{L^2}\right)}\left[\left(GJ + \frac{\pi^2 EI_w}{L^2} + \left\{\frac{\beta_x}{2}\sqrt{\left(\frac{\pi^2 EI_y}{L^2}\right)}\right\}^2\right)^{1/2} + \frac{\beta_x}{2}\sqrt{\left(\frac{\pi^2 EI_y}{L^2}\right)}\right] \tag{1.28}$$

1.3.2 Beams Carrying Transverse Loads

Comparatively little is known of the effects of variations in the conditions of loading and support on the lateral stability of monosymmetric beams. Anderson and Trahair (1972) have presented numerical results for the four cases: simply supported beams or cantilevers with either a concentrated load or a uniformly distributed load. These show that for simply supported beams the effects of flange loading and monosymmetry are as expected, i.e. the critical load increases both as the relative size of the top flange increases and/or as the load level relative to the shear centre decreases. Thus for top flange loading higher critical loads are always obtained when the larger flange is used as the compression flange. However, for cantilevers the two effects tend to oppose each other; for bottom flange loading higher values are obtained when this is the larger flange, whereas for loads applied above the level of the top flange, making this the larger flange may produce the higher buckling loads.

1.4 BUCKLING OF NON-UNIFORM BEAMS

1.4.1 General

For situations in which the pattern of major-axis moments produced by the applied loading varies in a regular manner over the span, the use of a beam of variable cross-section may well provide a structurally efficient solution. In simple cases it may be possible to match the variation in flexural rigidity about the major axis and the bending moment diagram quite closely.

1.4.2 Tapered Beams

Depth taper of narrow rectangular beams causes appreciable changes in both I_y and J leading to significant changes in lateral stability.

For flanged beams, however, depth variations have comparatively little effect on the beam's resistance to lateral buckling (except for sections which resist twisting largely through warping) since both I_y and J depend mainly upon the flange proportions. Any reductions in either flange breadth or flange thickness on the other hand produce corresponding reductions in lateral stability which are quite large. .

The buckling of beams provided with various types of linear taper, e.g. depth, flange width, combined depth and flange width, etc., has been studied by several authors (Krefield *et al.*, 1959; Massey and McGuire, 1971; Kitipornchai and Trahair, 1972, 1975; Nethercot, 1973*b*; Morrell and Lee, 1974). Both simply supported beams and cantilevers have been considered and results, often in numerical format, are available for several examples of this problem.

Based on the analysis of a number of different cross-sections provided with a variety of loading and support conditions, Nethercot (1973*b*) has suggested that the elastic critical load for a tapered beam may be closely approximated by applying the following reduction factor r to the elastic critical load calculated for an equivalent uniform beam possessing the properties of the cross-section at the point of maximum moment:

$$r = \frac{7 + \gamma}{5 + 3\gamma} \tag{1.29}$$

in which

$$\gamma = \frac{Z_{x0}}{Z_{x1}} \left[\left(\frac{D_1}{D_0}\right)^3 \left(\frac{B_0}{B_1}\right)^3 \left(\frac{T_0}{T_1}\right)^2 \right]$$

and subscripts 0 and 1 relate to the points of maximum and minimum moment respectively.

1.4.3 Stepped Beams

In practical terms it may be more attractive to vary section properties in discrete steps, e.g. by providing additional flange plates over parts of the span. The lateral buckling of this type of non-uniform beam has been studied by Trahair and Kitipornchai (1971) who have provided tabulated results for several examples.

1.5 ELASTIC BEHAVIOUR OF INITIALLY DEFORMED BEAMS

All of the foregoing discussion on the lateral–torsional buckling problem has been based on the notion of a perfect beam, i.e. it has assumed the member to be initially straight, to behave elastically and for the loading to cause pure major-axis bending. As in the case of struts, such a situation can only be approached but not attained in reality owing to the unavoidable presence in real structures of ‘imperfections’. A full treatment of all aspects of this topic as it affects the design of laterally unsupported beams is provided in Chapter 3. As part of the necessary preparation for this, the elastic behaviour of imperfect beams will be discussed herein; the effects of inelastic material behaviour and initial residual stresses are considered in Chapter 2.

The presence of either an initial lack of straightness (initial bow from the minor axis and/or initial twist) in the beam of Fig. 1.3 or of small eccentricities in the applied loading (producing small minor-axis moments) produces qualitatively similar effects on the beam's response. Taking the case of a beam containing an initial bow u_0 and an initial twist ϕ_0, eqns (1.1) and (1.2) must be modified to

$$EI_y \frac{d^2u}{dz^2} - M(\phi + \phi_0) \tag{1.30}$$

$$GJ\frac{d\phi}{dz} - EI_w \frac{d^3\phi}{dz^3} = M\left(\frac{du}{dz} - \frac{du_0}{dz}\right) \tag{1.31}$$

If u_0 and ϕ_0 are both assumed to adopt a half-sinewave shape (which approximates quite well to actual measurements), such that

$$u_0 = \delta_0 \sin \frac{\pi z}{L} \tag{1.32a}$$

$$\phi_0 = \theta_0 \sin \frac{\pi z}{L} \tag{1.32b}$$

and the central initial bow δ_0 and initial twist θ_0 are further assumed to be related by

$$\delta_0 = \theta_0 \frac{M_0}{\pi^2 EI_y/L^2} \tag{1.33}$$

then the additional lateral deflection u and twist ϕ at a moment M are given by

$$u = \frac{M/M_0}{1 - M/M_0} \delta_0 \sin \frac{\pi z}{L} \tag{1.34}$$

$$\phi = \frac{M/M_0}{1 - M/M_0} \theta_0 \sin \frac{\pi z}{L} \tag{1.35}$$

Figure 1.16 shows how u and ϕ increase with an increase in M, the rate of such increases accelerating as M approaches its upper limit of M_0.

Equation (1.34) may be used as a basis for determining stresses, the maximum value due to simultaneous major- and minor-axis bending and warping being given by

$$F_{max} = \frac{M}{Z_x} - \frac{EI_y}{Z_y}\left(\frac{d^2(u + (h/2)\phi)}{dz^2}\right)L/2 \tag{1.36}$$

If this is limited to the material yield stress F_Y, then eqn (1.36) may be rearranged to give the nominal applied bending stress $F_b = M/Z_x$ as the solution of

$$F_b = F_Y - \frac{\delta_0 P_y}{M_0}\left(1 + \frac{h}{2}\frac{P_y}{M_0}\right)\frac{Z_x}{Z_y}\left(\frac{F_b}{1 - F_b/F_{0b}}\right) \tag{1.37}$$

in which $P_y = \pi^2 EI_x/L^2$ and $F_b = M_0/Z_x$.

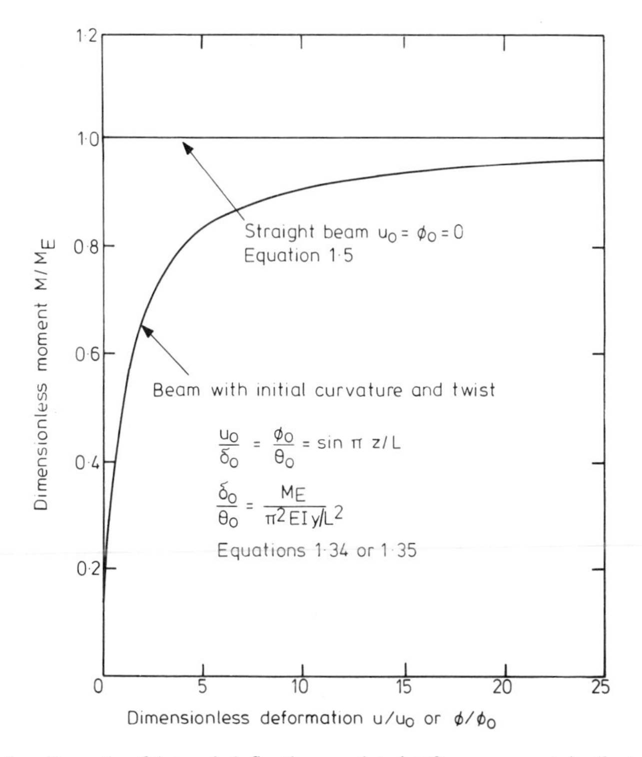

FIG. 1.16. Growth of lateral deflection and twist for a geometrically imperfect beam.

Figure 1.17 shows, for one particular value of δ_0, how these stresses are related to the elastic critical stress obtained from eqn (1.5). For short beams, loads which induce values of F_b approaching the yield stress may be applied, whilst first yield will be reached in slender beams when the applied loading is close to the elastic critical value.

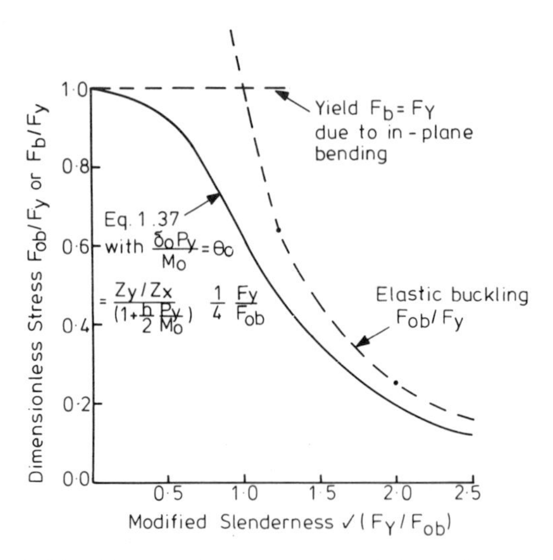

FIG. 1.17. Limiting stresses for initially imperfect beams.

1.6 CONCLUDING REMARKS

This first chapter has discussed the elastic lateral–torsional buckling of beams. Beginning with the general background in Section 1.1, it has proceeded to develop the basic theory in Section 1.2, after which it has covered various extensions and modifications in Sections 1.2–1.5. However, because this chapter has been written as the first of three devoted to the lateral buckling problem, it has not mentioned either inelastic behaviour or the design aspects of the subject. These topics are treated in Chapters 2 and 3 respectively, and the reader should now proceed to study these.

REFERENCES

AISC (1969) *Specification for the Design, Fabrication and Erection of Structural Steel for Buildings*, American Institute of Steel Construction, New York (Feb.).

ANDERSON, J. M. and TRAHAIR, N. S. (1972) Stability of monosymmetric beams and cantilevers. *Journal of the Structural Division, ASCE*, **98**(ST1), Proc. Paper 8646, 269–86.

AUSTIN, W. J., YEGIAN, S. and TUNG, T. P. (1955) Lateral buckling of elastically end-restrained beams. *Proceedings of the ASCE*, **81**, Separate No. 673, 1–25.

BARSOUM, R. S. and GALLAGHER, R. H. (1970) Finite element analysis of torsional and torsional-flexural stability problems. *International Journal of Numerical Methods in Engineering*, **2**, 335–52.

BARTELS, D. and BOS, C. A. M. (1973) Investigation of the effect of the boundary conditions on the lateral buckling phenomenon, taking account of cross-sectional deformation. *Heron*, **19**(1), 3–26.

BSI (1969) British Standard Code of Practice CP118:1969, *The Structural Use of Aluminium*, British Standards Institution, London.

BROWN, P. T. and TRAHAIR, N. S. (1975) Finite integral solution of torsion and plate problems. *Civil Engineering Transactions, Institution of Engineers, Australia*, **CE17**(2), 59–63.

CHEN, W. F. and ATSUTA, T. (1977) *Theory of Beam-Columns*, Vol. 2, McGraw-Hill, New York.

CUK, P. E. and TRAHAIR, N. A. (1981) Buckling of beams with concentrated moments. University of Sydney, School of Civil Engineering, Research Report No. R401 (Sept.).

DUX, P. F. and KITIPORNCHAI, S. (1978) Approximate inelastic buckling moments for determinate I-beams. *Civil Engineering Transactions, Institution of Engineers, Australia*, **CE20**(2), 128–33.

GALAMBOS, T. V. (1977) Laterally unsupported beams. *Introductory Report, 2nd International Colloquium on Stability of Steel Structures*, ECCS–IABSE, Liège, pp. 365–73.

HORNE, M. R. (1954) The flexural-torsional buckling of members of symmetrical I-section under combined thrust and unequal terminal moments. *Quarterly Journal of Mechanics and Applied Mathematics*, **7**(4), 410–26.

JOHNSTON, B. G. and EL DARWISH, I. A. (1965) Torsion of structural shapes. *Journal of the Structural Division, ASCE*, **91**(ST1), 203–28.

KIRBY, P. A. and NETHERCOT, D. A. (1979) *Design for Structural Stability*, Granada Publishing, St Albans.

KITIPORNCHAI, S. and TRAHAIR, N. S. (1972) Elastic stability of tapered I-beams. *Journal of the Structural Division, ASCE*, **98**(ST3), Proc. Paper 8775, 713–28.

KITIPORNCHAI, S. and TRAHAIR, N. S. (1975) Elastic behaviour of tapered monosymmetric I-beams. *Journal of the Structural Division, ASCE*, **101**(ST8), Proc. Paper 1149.

KITIPORNCHAI, S. and TRAHAIR, N. S. (1980) Buckling properties of monosymmetric I-beams. *Journal of the Structural Division, ASCE*, **106**(ST5), 941–57.

KREFIELD, W. J., BUTLER, D. J. and ANDERSON, G. B. (1959) Welded cantilever wedge beams. *Welding Research*, **38**, 97–112-s.

LEE, G. C. (1960) A survey of the literature on the lateral instability of beams. *Welding Research Council Bulletin*, No. 63 (Aug.).

MASSEY, C. and MCGUIRE, P. J. (1971) Lateral stability of nonuniform cantilevers. *Journal of the Engineering Mechanics Division, ASCE*, **97**(EM3), Proc. Paper 8172, 673–86.

MASUR, E. F. and MILBRADT, K. P. (1957) Collapse strength of redundant beams after lateral buckling. *Journal of Applied Mechanics, ASME*, **24**(2), 283.

MICHELL, A. G. M. (1899) Elastic stability of long beams under transverse forces. *Philosophical Mag.*, **48**, 298.

MORRELL, M. L. and LEE, G. C. (1974). Allowable stress for web-tapered beams. *Welding Research Council Bulletin*, No. 192 (Feb.), pp. 1–12.

NETHERCOT, D. A. (1972*a*) Recent progress in the application of the finite element method to problems of the lateral buckling of beams. *Proceedings, Conference on Finite Element Methods in Civil Engineering*, Engineering Institute of Canada, Montreal (June), pp. 367–91.

NETHERCOT, D. A. (1972*b*) The effect of load position on the lateral stability of beams. *Proceedings, Conference on Finite Element Methods in Civil Engineering*, Engineering Institute of Canada, Montreal (June), pp. 347–66.

NETHERCOT, D. A. (1972*c*) Influence of end support conditions on the stability of transversely loaded beams. *Building Science*, **7**, 87–94.

NETHERCOT, D. A. (1973*a*) The effective lengths of cantilevers as governed by lateral buckling. *Structural Engineer*, **51**(5), 161–8.

NETHERCOT, D. A. (1973*b*) Lateral buckling of tapered beams. *Publications, IABSE*, **33**(II), 173–92.

NETHERCOT, D. A. and ROCKEY, K. C. (1971) A unified approach to the elastic lateral buckling of beams. *Structural Engineer*, **49**(7), 321–30.

NETHERCOT, D. A. and ROCKEY, K. C. (1973) Lateral buckling of beams with mixed end conditions. *Structural Engineer*, **51**(4), 133–8.

NETHERCOT, D. A. and TRAHAIR, N. S. (1976) Lateral buckling approximations for elastic beams. *Structural Engineer*, **54**(6), 197–204.

NETHERCOT, D. A. and TRAHAIR, N. S. (1977) Lateral buckling calculations for braced beams. *Civil Engineering Transactions, Institution of Engineers, Australia*, **CE19**(2), 211–14.

PRANDTL, L. (1899) Kipperscheinungen. Dissertation, Munich.

SALVADORI, M. G. (1951) Lateral buckling of beams of rectangular cross-section under bending and shear. *Proceedings, 1st US National Congress of Applied Mechanics*, p. 403.

SALVADORI, M. G. (1955) Lateral buckling of I-beams. *Transactions, ASCE*, **120**, Paper 2773, 1165–77.

SALVADORI, M. G. (1956) Lateral buckling of eccentrically loaded I-columns. *Transactions, ASCE*, **121**, Paper 2836, 1163–78.

TERRINGTON, J. S. (1968). *Combined bending and torsion of beams and girders*. British Constructional Steelwork Association, Publication No. 31, Part 1.

TERRINGTON, J. S. (1970) Ibid., Part 2.

TIMOSHENKO, S. P. (1905) Einige Stabilitätsprobleme der Elasticitätstheorie. *Bulletin of the Polytechnic Institute, St Petersburg*; reprinted in *Zeitschrift für Mathematik und Physik*, **58** (1981), 337.

TIMOSHENKO, S. P. and GERE, J. M. (1961) *Theory of Elastic Stability*, 2nd Edn, McGraw-Hill, New York.

TRAHAIR, N. S. (1965) Stability of I-beams with elastic end restraints. *Journal of the Institution of Engineers, Australia*, **37**(6), 157–68.

TRAHAIR, N. S. (1968*a*) Elastic stability of propped cantilevers. *Civil Engineering Transactions, Institution of Engineers, Australia*, **CE10**(1), 94–100.

TRAHAIR, N. S. (1968*b*) Interaction buckling of narrow rectangular continuous beams. *Civil Engineering Transactions, Institution of Engineers, Australia*, **CE10**(2), 167–72.

TRAHAIR, N. S. (1969) Elastic stability of I-beam elements in rigid-jointed structures. *Journal of the Institution of Engineers, Australia*, **38**(7–8), 171–80.

TRAHAIR, N. S. (1977) Lateral buckling of beams and beam-columns. Chap. 3 of *Theory of Beam-Columns*, Vol. 2, ed. W. F. Chen and T. Atsuta, McGraw-Hill, New York, pp. 71–157.

TRAHAIR, N. S. and KITIPORNCHAI, S. (1971) Elastic lateral buckling of stepped I-beams. *Journal of the Structural Division, ASCE*, **97**(ST10), Proc. Paper 8445, 2535–48.

TRAHAIR, N. S. and WOOLCOCK, S. T. (1973) Effect of major axis curvature on I-beam stability. *Journal of the Engineering Mechanics Division, ASCE*, **99**(EM1), Proc. Paper 9548, 85–98.

VACHARAJITTIPHAN, P. and TRAHAIR, N. S. (1974) Direct stiffness analysis of lateral buckling. *Journal of Structural Mechanics*, **3**(2), 107–37.

VACHARAJITTIPHAN, P., WOOLCOCK, S. T. and TRAHAIR, N. A. (1974) Effect of in-plane deformation on lateral buckling. *Journal of Structural Mechanics*, **3**(1), 29–60.

VINNAKOTA, S. (1977) Finite difference method for plastic beam-columns. Chap. 10 of *Theory of Beam-Columns*, Vol. 2, ed. W. F. Chen and T. Atsuta, McGraw-Hill, New York, pp. 451–503.

WOOLCOCK, S. T. and TRAHAIR, N. S. (1974) The post-buckling behaviour of determinate beams. *Journal of the Engineering Mechanics Division, ASCE*, **100**(EM3), 151–72.

WOOLCOCK, S. T. and TRAHAIR, N. S. (1976) The post-buckling of redundant I-beams. *Journal of the Engineering Mechanics Division, ASCE*, **102**(EM2), 293–312.

ZBIROHOWSKI-KOSCIA, K. (1967) *Thin Walled Beams*, Crosby Lockwood, London.

Chapter 2

INELASTIC LATERAL BUCKLING OF BEAMS

N. S. Trahair

School of Civil and Mining Engineering, University of Sydney, New South Wales, Australia

SUMMARY

The inelastic lateral buckling of steel beams is discussed. A theoretical model is presented for predicting the buckling strength reductions caused by yielding, and methods of solution are summarised. The effects of residual stresses due to hot-rolling, flame-cutting and welding are discussed, and theoretical predictions are presented for a range of beam support and loading conditions.

NOTATION

a	Distance of load from support
a_{1-6}	Coefficients defining quartic residual stress distributions
$\bar{b}$	Height of load above geometrical axis
f	Stress
f_{rc}	Residual compressive stress in welded flange
f_{rf}, f_{rw}	Residual stresses in flange and web
f_{rft}, f_{rfw} and f_{rwc}	Residual stresses at flange tip, flange centre and web centre
h	Distance between flange centroids
r_y	Minor-axis radius of gyration
t	Web thickness
u	Deflection of shear centre in x-direction

u_0	Geometrical imperfection
v	Deflection of shear centre in y-direction
x, y	Major and minor principal axes of section
$\bar{y}$	Distance from flange centroid to section centroid
y_n	Coordinate of inelastic neutral axis
y_0	Coordinate of shear centre
y_{0t}	Coordinate of tangent modulus shear centre
z	Longitudinal axis through centroid
A	Cross-sectional area, or matrix (eqn (2.16))
A_w	Area of weld metal
B	Flange width
B_B, B_T	Width of bottom and top flanges
B_w	Energy supplied per unit volume of electrode wire
D	Depth of section
E	Young's modulus of elasticity
E_s	Strain-hardening modulus
E_t	Tangent modulus
F_Y	Yield stress
G	Shear modulus of elasticity, or stability matrix (eqn (2.18))
G_s	Strain-hardening shear modulus
G_t	Tangent shear modulus
I_w	Warping section constant
I_x, I_y	Second moments of area about the x, y axes
J	Torsion section constant
K	Stiffness matrix (eqn (2.18))
L	Span or segment length
M	Moment
M_c	Buckling moment
M_E	Elastic buckling moment
M_I	Inelastic buckling moment
M_P	Full plastic moment
M_x	Major-axis bending moment
M_Y	Nominal first yield moment $= F_Y Z_x$
P	Transverse concentrated load, or axial load on compression member
P_E	Elastic buckling load
P_I	Inelastic buckling load
P_r	Reduced modulus compression load

P_t	Tangent modulus compression load
P_Y	Load at nominal first yield
T	Flange thickness
T_B, T_T	Thickness of bottom and top flanges
Z_x	Major-axis elastic section modulus
β	Ratio of end moments
β_x	Monosymmetry section constant
γ	Coefficient defining value of end moment
γ_L, γ_R	Values of γ for left and right ends
ε	Strain
ε_s	Strain at beginning of strain-hardening
ε_Y	Yield strain
λ	Load factor
ν	Poisson's ratio
ϕ	Angle of twist
$(\ )_t$	Tangent modulus value
$'$	$\equiv d/dz$

2.1 INTRODUCTION

The lateral buckling of a perfectly straight elastic beam loaded in its stiffer principal plane (Fig. 2.1) is fully discussed in Chapter 1. There it is shown that if a beam has low resistances to lateral bending and torsion, then it may buckle suddenly by deflecting and twisting out of the plane of loading.

The results obtained in Chapter 1 are only valid while the beam remains elastic, which is the case for a slender long-span beam. For a short-span steel beam, the resistance to elastic buckling is high, and yielding may commence before the elastic buckling load can be reached. Yielding substantially reduces the resistance of a beam to buckling, and the inelastic buckling load may be much lower than the elastic buckling load, as shown in Fig. 2.2. Yielding of a straight beam is affected by its in-plane loading and by any residual stresses (Fig. 2.3) that are induced during manufacture. A theoretical model of the inelastic buckling of steel beams is developed in Section 2.2, and methods of analysing this model are compared in Section 2.3. The application of these analyses to a range of beam support and loading conditions is discussed in Section 2.4.

The ultimate strength of a short-span beam is reduced below its inelastic

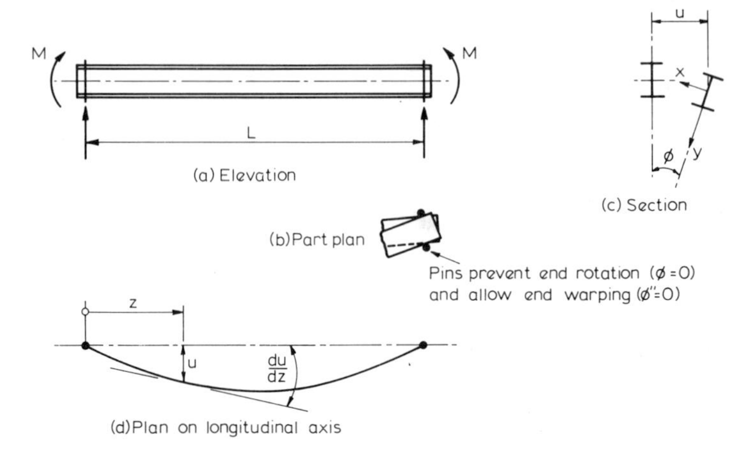

FIG. 2.1. Buckling of a simply supported I-beam.

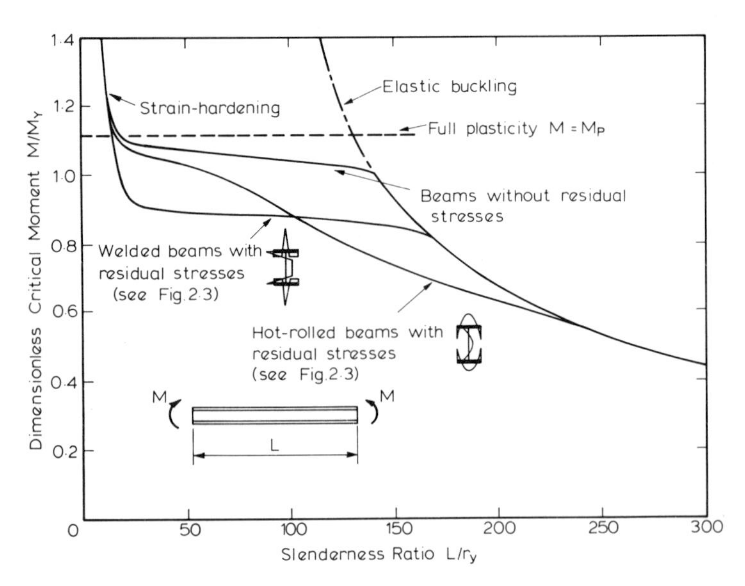

FIG. 2.2. Lateral buckling strengths of simply supported I-beams.

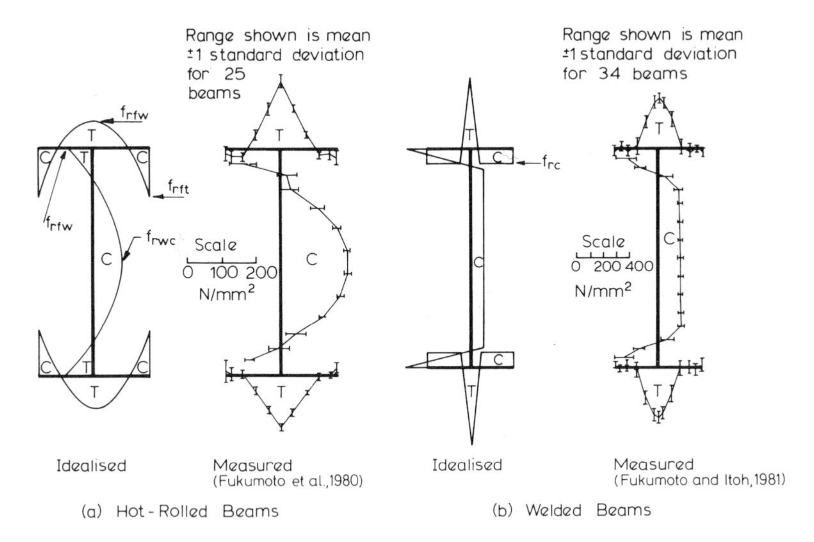

FIG. 2.3. Residual stress patterns.

buckling strength by the presence of geometrical imperfections, such as initial crookedness and twist, which occur in real beams. The results of experimental studies of the ultimate strengths of beams are compared with inelastic buckling predictions in Chapter 3, and design methods are discussed which take into account the influence of inelastic buckling and geometrical imperfections.

2.2 MODELS FOR INELASTIC BUCKLING

2.2.1 General

The elastic lateral buckling of a three-dimensional beam is usually analysed by modelling it either as an elastic line, or as a series of plate elements. For the elastic line model, the three-dimensional nature of the beam is allowed for by positioning the line at the shear centre axis, and by using the cross-section properties EI_x, EI_y, GJ, EI_w and β_x in the analysis of the in-plane bending, lateral bending and torsion of the elastic line. For the plate element model, the three-dimensional behaviour of the beam is synthesised from the in-plane and out-of-plane behaviour of the individual flat plate elements into which the beam is divided.

The elastic line model may be adapted in two stages to represent the lateral buckling of an inelastic steel beam. In the first stage, the pre-buckling in-plane bending is analysed, first by a cross-section analysis for the variations of the yield and strain-hardening boundaries with moment, and then by a beam analysis to determine the spatial distributions of these boundaries along the beam. Data for and details of the cross-section analysis are given in Section 2.2.2, and a method of beam analysis in Section 2.3.1.

For the second stage, a model of the out-of-plane bending and torsion of the line is developed in Section 2.2.3 by accounting for the cross-sectional distributions of the yield and strain-hardening regions, by using tangent moduli of elasticity E_t, G_t to calculate the effective values of the cross-section properties, and by modifying the elastic equations of bending and torsion of uniform beams to allow for the variation of these properties along the beam. The analysis of this inelastic line model is discussed in Section 2.3.2.

Alternatively, a similar process might be used to calculate the in-plane and out-of-plane properties of each flat plate in the beam, and to synthesise the out-of-plane behaviour of the beam.

2.2.2 In-Plane Bending of Cross-Section

(a) Section Geometry and Material Properties

The geometry of an idealised doubly symmetric I-beam is defined in Fig. 2.4(a). For simplicity, all radii and fillets are ignored, as is any flange tapering.

An idealised tensile stress–strain relationship for structural steel is shown

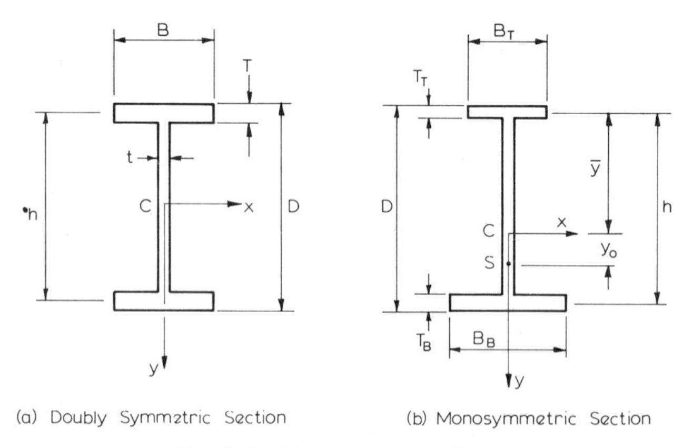

FIG. 2.4. Beam cross-sections.

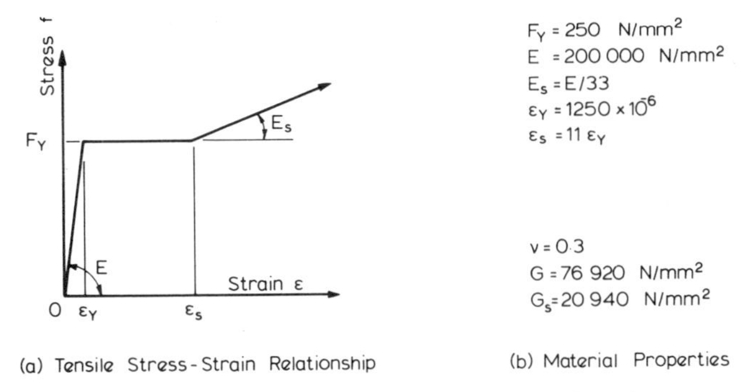

FIG. 2.5. Idealised properties of structural steel.

in Fig. 2.5, with nominal values for E, E_s, F_Y and ε_s. The existence of an upper yield stress is ignored. This stress–strain curve is assumed to apply to all of the material in the steel beams considered. It may be noted that the flange yield stress of a real beam is often higher than the nominal value, and that the web yield stress is usually significantly greater than the flange yield stress.

Values of the shear moduli G and G_s are also given in Fig. 2.5. The elastic shear modulus G is determined from

$$G = E/2(1 + \nu) \tag{2.1}$$

by using a Poisson's ratio of $\nu = 0{\cdot}3$. The strain-hardening shear modulus G_s is approximated from (Lay, 1965)

$$G_s/G = 8/\{4 + (E/E_s)/(1 + \nu)\} \tag{2.2}$$

(*b*) *Residual Stresses in Hot-Rolled Beams*

Longitudinal residual stresses are induced in a hot-rolled steel beam during cooling after rolling, and as a result of any mechanical straightening processes (Lay and Ward, 1969). During cooling, the more highly exposed regions of the cross-section at the flange tips and web centre cool more rapidly. These early cooling regions shrink, inducing matching plastic flows in the high temperature late-cooling regions at the flange–web junctions, which have correspondingly low yield stresses. Subsequent shrinkages of the high temperature regions during final cooling are resisted by the regions already cooled which have developed high yield stresses, and induce residual compressions in them. Equilibrating residual tensions are induced in the late-cooling regions. Cold-working by mechanical straightening causes local yielding, and further modifies the residual stress pattern.

The magnitudes and distributions of residual stress vary considerably with the cross-section geometry and with the cooling and straightening processes. Idealised and measured distributions of residual stress in hot-rolled beams are shown in Fig. 2.3(a). The distributions across flange and web are often approximately parabolic, being compressive at the early-cooling flange tips and web centre, and tensile at the late-cooling flange–web junctions. Young (1975) has suggested that the maximum residual stresses in hot-rolled beams may be approximated by

$$\begin{aligned} f_{\mathrm{rft}} &= 137{\cdot}5(2{\cdot}2 - A/2BT)\ \mathrm{N\,mm^{-2}} \\ f_{\mathrm{rfw}} &= 100(-0{\cdot}3 + A/2BT)\ \mathrm{N\,mm^{-2}} \\ f_{\mathrm{rwc}} &= 83{\cdot}3(0{\cdot}8 + A/2BT)\ \mathrm{N\,mm^{-2}} \end{aligned} \tag{2.3}$$

While the flange and web distributions are approximately parabolic, it has been suggested (Lee *et al.*, 1967) that quartic distributions of the form

$$\begin{aligned} f_{\mathrm{rf}} &= a_1 + a_2 x^2 + a_3 x^4 \\ f_{\mathrm{rw}} &= a_4 + a_5 y^2 + a_6 y^4 \end{aligned} \tag{2.4}$$

should be used, so that the conditions

$$\int_A f_{\mathrm{r}}\,\mathrm{d}A = 0 \tag{2.5}$$

$$\int_A f_{\mathrm{r}}(x^2 + y^2)\,\mathrm{d}A = 0 \tag{2.6}$$

are satisfied. The first of these conditions ensures that the residual stresses have a zero axial force resultant (zero moment resultants about the x and y axes are automatically ensured by symmetry). The second condition was suggested to ensure that the residual stresses have a zero axial torque effect when the member is twisted elastically. If this condition is not satisfied, then the effective elastic torsional rigidity of the cross-section changes to $(GJ - \int_A f_{\mathrm{r}}(x^2 + y^2)\,\mathrm{d}A)$.

In practice, some residual stress distributions may differ significantly from the idealised pattern discussed above, as can be seen from the measured distributions shown in Fig. 2.3(a) (Fukumoto *et al.*, 1980).

(*c*) *Residual Stresses in Welded Beams*

Residual stresses are induced in welded beams by uneven heating and cooling as a result of flame-cutting and welding the flanges and webs. The process is similar to that described above for hot-rolled beams, with large residual tensions being induced in the late-cooling flange–web junctions, and equilibrating compressions in the flanges and web, as shown in Fig. 2.3(b).

The magnitudes and distributions of residual stress again vary with the cross-section geometry and with the cutting, welding, cooling and straightening processes. In the idealised residual stress distribution shown in Fig. 2.3(b), the compressive stresses are uniform across the flanges and web, except near the flange–web junctions. Dwight and White (1977) have suggested that the maximum tensile stress can be assumed to be equal to the parent metal yield stress, and that the compressive stresses can be estimated from

$$f_{rc} = 0{\cdot}2B_w \sum (A_w/A) \tag{2.7}$$

in which A_w is the area of weld metal added in the biggest single pass at the weld site, by using their suggested values of B_w, the energy supplied per unit volume of electrode wire.

Again, practical residual stress distributions may differ significantly from this idealised pattern, as can be seen from the measured distributions shown in Fig. 2.3(b) (Fukumoto and Itoh, 1981).

(d) Cross-Section Analysis of Yielding

The in-plane bending of a beam with residual stresses may be initially elastic, but sooner or later yielding will commence at the most highly strained regions, and spread progressively through the cross-section. At high moments, the strain-hardening strain ε_s may be exceeded, in which case strain-hardening regions will develop in which the yield stress F_Y is exceeded.

Theoretically, yielding is caused by a combination of longitudinal normal stresses with shear stresses, these stresses being both residual and induced by bending and shear actions. For simplicity, it has become customary to ignore the effect of shear stresses on the yielding of beams which may buckle inelastically.

In this case, yielding can be determined by comparing the total strain

$$\varepsilon = (y_n - y)v'' + f_r/E \tag{2.8}$$

in which y_n defines the neutral axis position and v'' is the in-plane curvature, with the yield strain

$$\varepsilon_Y = F_Y/E \tag{2.9}$$

as shown in Fig. 2.6. It can be seen that the total strain distribution is monosymmetric, and that the neutral axis moves y_n from the geometrical axis at the web centreline. The position of the neutral axis is such that the axial force resultant of the total stresses f calculated from the total strains ε and the stress–strain relationship (Fig. 2.5) is zero, i.e.

$$\int_A f \, dA = 0 \tag{2.10}$$

This condition allows a value of y_n to be determined iteratively for each value of the in-plane curvature.

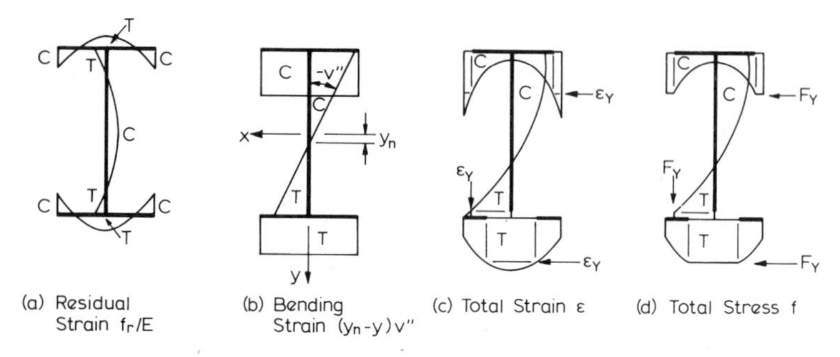

FIG. 2.6. Strain and stress distributions.

The in-plane bending moment M_x can then be determined as the moment resultant of the stresses f, i.e.

$$M_x = \int_A f y \, dA \tag{2.11}$$

A typical variation of M_x with curvature v'' is shown in Fig. 2.7.

The boundaries of the yielded or strain-hardened regions can be determined as the positions in the cross-section where the total strain equals the yield or strain-hardening strains ε_Y or ε_s. Typical variations of these boundaries with the major-axis moment M_x are shown in Fig. 2.8 for the case where the residual compressive stress at the flange tip is equal to $0{\cdot}5F_Y$. The top flange commences yielding at the flange tip at $0{\cdot}5M_Y$ ($M_Y = F_Y Z_x$ is the nominal first yield moment in the absence of residual stresses), and is fully plastic at $1{\cdot}09M_Y$. The bottom flange commences yielding at $0{\cdot}7M_Y$ at its centre (where the residual tensile stress was $0{\cdot}3F_Y$), and is fully plastic at

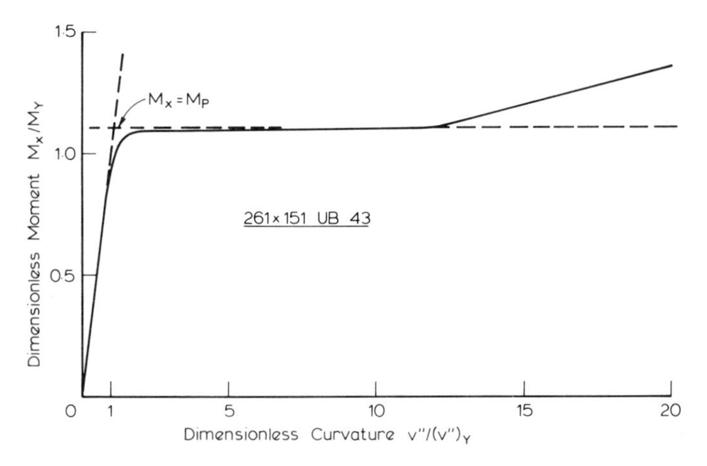

FIG. 2.7. Moment–curvature relationship.

$1{\cdot}08M_Y$. Yielding of the web commences at the bottom at $0{\cdot}76M_Y$, and progresses slowly until the flanges are fully yielded at $1{\cdot}09M_Y$. The process then accelerates, and the web is virtually fully yielded at $1{\cdot}14M_Y$, when strain-hardening commences in both flanges.

The position of the neutral axis y_n is also shown in Fig. 2.8. As yielding progresses, the neutral axis moves slightly away from the geometrical axis through the web centreline, but returns to the web centreline after both flanges are fully yielded.

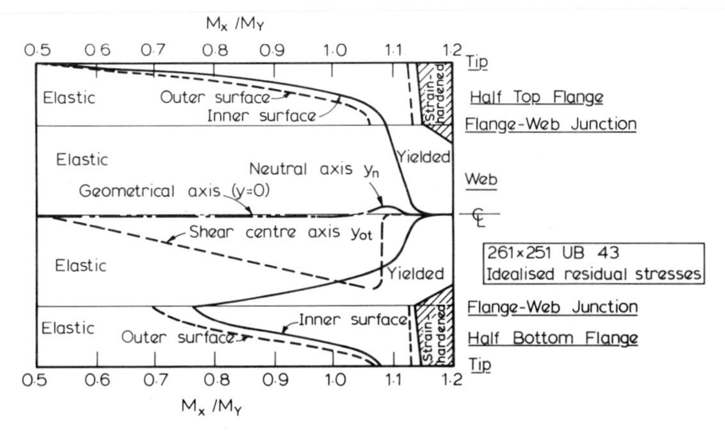

FIG. 2.8. Propagation of yielded boundaries.

2.2.3 Line Model for Inelastic Buckling

(a) Tangent Modulus Theory of Inelastic Buckling

Most inelastic column buckling studies are based on the tangent modulus theory (Trahair, 1977), in which the modulus used to estimate the resistance to buckling is the tangent to the stress–strain curve at the particular stress level considered, as shown in Fig. 2.9(a). Although this theory appears to be invalid for inelastic materials which unload elastically (Fig. 2.9(b)), careful experiments have shown that it leads to more accurate predictions than the apparently rigorous reduced modulus theory (Trahair, 1977) which accounts for elastic unloading. This paradox was resolved by Shanley (1947), who reasoned that the tangent modulus theory is valid when buckling is accompanied by a simultaneous increase in the applied load of sufficient magnitude to prevent unloading (Fig. 2.9(c)). Thus, while the reduced modulus load P_r is the highest load under which the inelastic member can remain straight, the tangent modulus load P_t is the lowest load at which inelastic buckling can begin.

The reasons for the wide acceptance of the tangent modulus theory for inelastic column buckling are that it is simpler to use than the reduced modulus theory, and that it provides a conservative estimate of the member strength which is in closer agreement with experimental results. It is also generally accepted as the basis for inelastic beam buckling theories.

The application of the tangent modulus theory to the idealised stress–strain curve shown in Fig. 2.5(a) leads to the use of E (and G) for the elastic regions, and E_s (and G_s) for the strain-hardened regions, and suggests

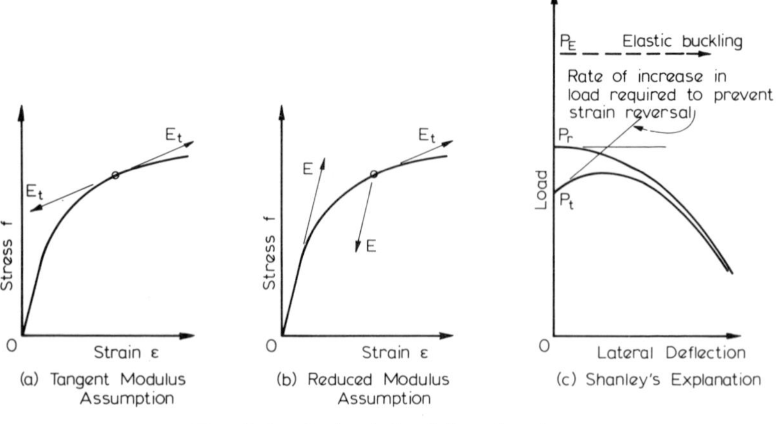

FIG. 2.9. Inelastic buckling theories.

$E_t = 0$ (and $G_t = 0$) for the yielded regions. However, experimental evidence indicates that zero values are too low for the tangent moduli for the yielded regions. White (1960) suggested that yielding may be considered as a series of discontinuous slips, and that all of the material in the yielded region is either elastic or strain-hardened. This led to the proposal that the tangent moduli for the yielded regions should conservatively be taken as equal to the strain-hardening values E_s and G_s.

(*b*) *Inelastic Section Rigidities*

The resistance to elastic buckling (Chapter 1) is generated by the minor-axis flexural rigidity EI_y, the torsional rigidity GJ and the warping rigidity EI_w. Also important are the distances $(\bar{b} + y_0)$ of any loads above the axis of twist through the shear centre, and the section property β_x (Chapter 1), which reflects the influence of the longitudinal stresses on the effective torsional rigidity $(GJ + M_x\beta_x)$ of a monosymmetric I-beam. For the tangent modulus theory of inelastic buckling, new values must be calculated for these quantities which are appropriate for the inelastic cross-section.

The tangent flexural rigidity $(EI_y)_t$ of the section can be simply calculated (Trahair and Kitipornchai, 1972), as can the tangent torsional rigidity $(GJ)_t$ if each flange or web is idealised as a bimetallic strip (Booker and Kitipornchai, 1971). The variations of $(EI_y)_t$ and $(GJ)_t$ with moment M_x are shown non-dimensionally in Fig. 2.10 (Kitipornchai and Trahair, 1975*a*). These decrease steadily after yielding commences at $0{\cdot}5M_Y$, then more rapidly as the flanges approach full plasticity, and finally attain steady values when the flanges are fully yielded.

The shear centre axis position y_{0t} in an inelastic beam differs from the geometric axis $(y = 0)$ and the neutral axis (y_n) because of the monosymmetry of the inelastic cross-section. Trahair and Kitipornchai (1972) have shown how to calculate the position of the shear centre axis in an inelastic beam, and have demonstrated the variation of this with the major-axis moment M_x, as shown in Fig. 2.8 (Kitipornchai and Trahair, 1975*a*). The shear centre axis moves towards the bottom flange as M_x increases from $0{\cdot}5M_Y$ to $1{\cdot}08M_Y$. This happens because the bottom flange is the more effective as a result of the significant reductions in the effective rigidity of the top flange caused by early yielding near the flange tips (yielding at the centre of the bottom flange causes only minor reductions in its rigidity). For $M_x > 1{\cdot}08M_Y$, both flanges are completely yielded or strain-hardened, and so the effective section is doubly symmetric again, and the shear centre moves back to the geometrical axis.

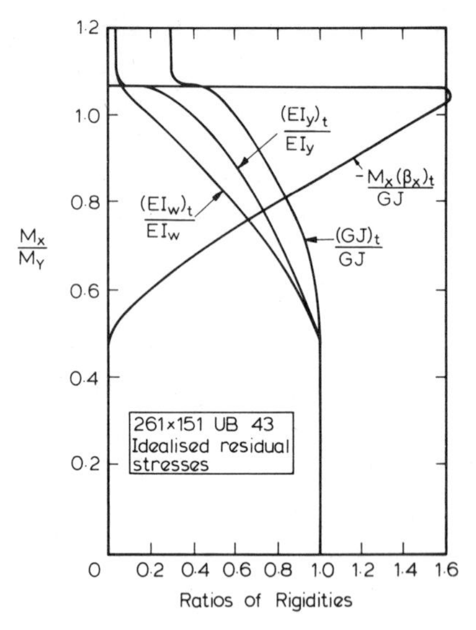

FIG. 2.10. Variations of the inelastic rigidities with moment.

The tangent warping rigidity $(EI_w)_t$ may be calculated from the effective flange flexural rigidities and the position of the shear centre (Trahair and Kitipornchai, 1972) by using

$$(EI_w)_t = (EI_y)_t(h^2/4)(1 - 4y_{0t}^2/h^2) \tag{2.12}$$

The variation of $(EI_w)_t$ with M_x shown in Fig. 2.10 is similar to that of $(EI_y)_t$.

Because of the monosymmetry of the yielded cross-section and the movement of the shear centre axis away from the geometrical axis, the longitudinal bending stresses due to the moment M_x will exert a torque during twisting which will change the effective torsional rigidity to $((GJ)_t + M_x(\beta_x)_t)$, in which

$$M_x(\beta_x)_t = \int_A f(x^2 + (y - y_{0t})^2)\,\mathrm{d}A \tag{2.13}$$

The variation of $-M_x(\beta_x)_t$ with M_x is shown non-dimensionally in Fig. 2.10. This remains zero until yielding commences at $0{\cdot}5M_Y$, and then increases steadily as M_x increases towards $1{\cdot}08M_Y$ and the shear centre axis moves towards the bottom flange. For $M_x > 1{\cdot}08M_Y$ both flanges are fully yielded or strain-hardened and the section is doubly symmetric again, so $-M_x(\beta_x)_t$ reduces to zero.

(*c*) *Non-Uniform Yielding*

The usual methods of analysing the minor-axis bending and torsion of elastic beams are for beams that have constant properties. However, when the bending moment M_x varies along the length of an inelastic beam, so do the effective cross-section and its properties, and the beam is non-uniform. Because of this, the minor-axis bending and torsion analyses must be modified to allow for this non-uniformity, or approximations must be developed which will permit the use of uniform beam theories.

Kitipornchai and Trahair (1975*c*) developed a bending and torsion theory for analysing the elastic buckling of tapered monosymmetric I-beams, and adapted this to study the inelastic buckling of the simply supported beam with central concentrated load shown in Fig. 2.11 (Kitipornchai and Trahair, 1975*a*). For the left-hand half of this beam, the minor-axis bending and torsion equations are

$$(EI_y)_t \frac{d^2u}{dz^2} + 2(EI_y)_t \frac{dy_{0t}}{dz}\frac{d\phi}{dz} = -M_x\phi \tag{2.14}$$

$$-3M_x\phi \frac{dy_{0t}}{dz} + [(GJ)_t + M_x(\beta_x)_t]\frac{d\phi}{dz} - \frac{d}{dz}\left[(EI_w)_t \frac{d^2\phi}{dz^2}\right]$$
$$= \frac{dM_x}{dz}[(u + (\bar{b} + y_{0t})\phi)_{L/2} - u] + M_x \frac{du}{dz} \tag{2.15}$$

in which u is the lateral deflection of the shear centre and ϕ is the angle of twist. They noted that special account had to be taken when solving these equations of any step or slope discontinuities of the shear centre axis.

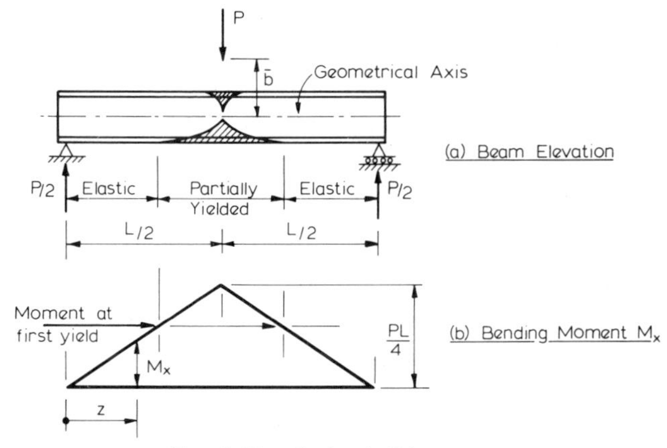

FIG. 2.11. Inelastic I-beam.

2.3 METHODS OF ANALYSIS

2.3.1 In-Plane Bending Analysis

The distribution of yielding along a beam due to in-plane bending and residual stresses can be determined by an in-plane bending analysis. From this, the variation along the beam of the out-of-plane buckling section properties can be determined. In general, these are calculated at a number of predetermined positions along the length of the beam.

The first step in the in-plane analysis is to determine the relationship between the in-plane bending moment M_x acting at a section and the curvature v'', and to find the positions in the cross-section of the yielded and strain-hardened boundaries. A method for doing this is described above in Section 2.2.2(d) and demonstrated in Figs. 2.6, 2.7 and 2.8. This also allows the variations with M_x of the out-of-plane buckling section properties to be determined, as described above in Section 2.2.3(b) and shown in Figs. 2.8 and 2.10.

When the beam is statically determinate in-plane, no further in-plane analysis is required, since in this case the spatial distribution of the major-axis moment M_x can be determined from statics alone. This then allows the immediate determination of the spatial distributions of the out-of-plane buckling properties.

When the beam is not statically determinate in-plane, then the spatial distribution of M_x cannot be determined from statics alone. In this case, the major-axis moment–curvature relationship must be integrated iteratively.

When there is only one redundant in-plane action, then the following trial-and-error method can be used:

1. Determine a trial value of the redundant action, based either on a previous analysis, or else on an interpolation between the elastic and fully plastic values.
2. Determine the spatial moment distribution, and from this the spatial curvature distribution.
3. Integrate numerically the curvature distribution, leaving the constants of integration undetermined.
4. Determine these constants of integration by satisfying the corresponding constraint conditions.
5. Evaluate the deflection or slope associated with the unsatisfied constraint imposed by the redundant action.
6. Repeat steps 2 to 5 for an increased value of the redundant action.
7. Extrapolate to obtain an improved value of the redundant action which will more nearly satisfy its constraint.

8. Repeat steps 2 to 5 and 7, until the constraint condition is sufficiently well satisfied.
9. Use the final spatial moment distribution to determine the spatial distributions of the out-of-plane buckling properties.

If there is a number of redundant in-plane actions, there will be a corresponding number of constraint conditions imposed by these redundants which must be satisfied. In this case a set of trial values of the redundants may be assumed and increased, and the corresponding changes in the unsatisfied constraint deformations calculated. This information can then be used to calculate new values of the redundants which will more nearly satisfy the constraints, and the process can be repeated until the desired convergence is achieved.

2.3.2 Out-of-Plane Buckling Analysis

(a) General

In Section 2.2.3 a model was developed for inelastic beam buckling in which the yielded and strain-hardened regions of an inelastic beam are taken to have elastic moduli equal to the strain-hardening values E_s and G_s. This produces a quasi-elastic model for the out-of-plane buckling, and any convenient elastic method of lateral buckling analysis can be used to determine the buckling load of this model.

The usual elastic methods of buckling analysis represent the three-dimensional beam as a one-dimensional line, and make use of the effective cross-section properties $(EI_y)_t$, $(GJ)_t$, $(EI_w)_t$, $(\beta_x)_t$, etc. Usually, they lead to the development of a set of linear equations of the type

$$[A]\{u\} = \{0\} \tag{2.16}$$

in which the vector $\{u\}$ is a representative set of the out-of-plane displacements u, ϕ or their derivatives at nodes along the beam, and the matrix $[A]$ contains the effective cross-section properties (which for inelastic beams depend on the load level) and terms representing the destabilising effects of the applied loads. Non-trivial solutions of these equations are obtained by finding the load levels for which

$$|A| = 0 \tag{2.17}$$

which define a series of buckling loads and a corresponding series of buckled shapes obtained from the vectors $\{u\}$.

For some elastic methods of buckling analysis, eqn (2.16) can be developed in the form

$$([K] - \lambda[G])\{u\} = \{0\} \tag{2.18}$$

in which $[K]$ is the stiffness matrix which depends on the effective cross-section properties, $[G]$ is the stability matrix representing the destabilising effects of an initial set of applied loads, and λ is the load factor. For elastic problems, $[K]$ and $[G]$ are independent of the loads, and eqn (2.18) is a standard eigenvalue equation for which efficient methods have been developed (Bishop *et al.*, 1965; Wilkinson, 1960) for the extraction of the buckling load factors λ and the eigenvectors $\{u\}$ (defining the buckled shapes).

However, in inelastic buckling problems, the stiffness and stability matrices $[K]$ and $[G]$ must be recalculated for each load level because of the changes in the inelastic section properties, and so these eigenvalue extraction methods lose their efficiency. In general, a more reasonable computation procedure is to iterate through a series of load levels towards a solution. At each load level the in-plane analysis is performed, and the spatial distribution of the out-of-plane section properties is determined. These are then used to establish the matrix $[A]$, and the value of its determinant $|A|$ is calculated. The iterations are continued until a zero determinant is calculated, which defines a buckling load, and enables a corresponding buckling mode $\{u\}$ to be determined. Some care must be taken with this method to ensure that the lowest buckling load is not missed. This can usually be done by steadily increasing the load level from first yield by reasonably small increments.

Computer methods of obtaining the matrix $[A]$ may be based on the differential equations which govern the out-of-plane bending and twisting of the line model of the beam, or on the balance between strain energy stored and work done during buckling. They include the methods of finite differences, finite integrals, transfer matrices and finite elements. These are discussed later in Sections 2.3.2(b)–(e).

The line methods of elastic buckling analysis are based on certain assumptions concerning the out-of-plane bending and torsional behaviour of a uniform monosymmetric beam, particularly the assumption that the axis of twist passes through the shear centre, which allows the decoupling of the out-of-plane bending and torsional resistances. The validity of applying these assumptions to a non-uniform monosymmetric beam is by no means universally accepted. Instead, it is common to make approximate analyses which either ignore the non-uniform nature of the inelastic beam, or approximate it by using a series of uniform elements to represent the beam.

This difficulty might be avoided by using a plate assembly model of an inelastic beam, in which the flanges and web are regarded as flat inelastic plates whose displacements are related to the web centreline displacements and the twist rotations of the cross-section. The in-plane and out-of-plane behaviours of these inelastic plates may then be analysed to synthesise the behaviour of the inelastic beam. Such finite plate (or strip) methods have been used to analyse the interaction between lateral buckling and local or distortional buckling in elastic beams whose cross-sections may distort (Johnson and Will, 1974; Rajasekaran, 1977; Hancock, 1978; Bradford and Trahair, 1981). When cross-section distortion is suppressed, these methods give results for uniform beams which are virtually identical to those obtained from the elastic line methods, although at some computational expense. There seems no reason why this method (with distortion suppressed) should not be adapted to the lateral buckling of inelastic beams.

(b) Finite Difference Method

In this method, finite differences (Collatz, 1966) are used to represent the higher-order differential coefficients such as u'' and ϕ''' at each of a number of nodes along a beam by linear combinations of the values of u, ϕ at these and adjacent nodes. This allows the continuous differential equations of out-of-plane bending and torsion (such as eqns (2.14) and (2.15)) to be replaced by a series of linear equations of the form of eqn (2.16) or (2.18), one for each node along the beam. This method has been used by a number of investigators to study inelastic buckling, including White (1960), Fukumoto and Galambos (1966), Abdel-Sayed and Aglan (1973), Yoshida and Nishida (1972), Djalaly (1974*a*, *b*) and Jarnot and Young (1977).

Vinnakota (1977) used the method in a slightly different way to study the inelastic biaxial bending of beams with small geometrical imperfections $\{u_0\}$. He developed a set of linear equations

$$[A]\{u\} = \{u_0\} \tag{2.19}$$

which are identical to eqn (2.16) if the beam is perfectly straight ($\{u_0\} = \{0\}$). He solved his equations iteratively for a series of increasing loads until the solution vector $\{u\}$ began to diverge, indicating that the maximum load capacity P_u had been reached.

Vinnakota (1977) compared his theoretical calculations for hot-rolled simply supported beams with unbraced central concentrated loads (Fig. 2.11) with those derived by Kitipornchai and Trahair (1975*a*) using the finite integral method (Fig. 2.12(a)). The only data difference reported by Vinnakota is that he assumed a sinusoidal initial crookedness with a maximum value of $L/1000$, instead of the straight beam assumed by Kitipornchai and Trahair. The two sets of results shown in Fig. 2.12(a) are in reasonable agreement, considering the different imperfections and approaches to non-uniformity and yielding used in the two investigations.

Vinnakota (1977) also compared his theoretical calculations for a set of four slightly different hot-rolled beams which had been tested by Kitipornchai and Trahair (1975*b*). Most of his theoretical predictions (Fig. 2.12(b)) are significantly lower than theirs, because he assumed an idealised residual stress pattern with comparatively high stresses at the flange tips instead of the low values measured. However, some of his predictions are quite close to the experimental failure loads.

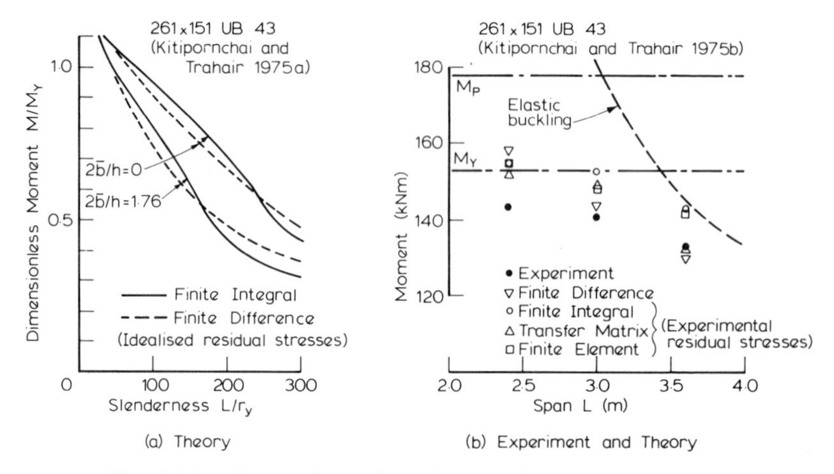

FIG. 2.12. Comparison of simple beam buckling moments.

In another study, Vinnakota (1977) compared his theoretical predictions for hot-rolled simply supported beams with unbraced central concentrated loads (Fig. 2.11) with those derived by Yoshida and Imoto (1973) using the transfer matrix method (Fig. 2.13). Vinnakota attributed the differences between the two sets of predictions not only to the different imperfections, but also to different allowances being made for the effects of yielding on the shear centre position, and of pre-buckling deflections. It is also probable that there are differences in the treatment of non-uniformity in the two methods of analysis.

(*c*) *Finite Integral Method*

The finite integral method (Brown and Trahair, 1968) is a reversal of the finite difference method, in that it represents the lower-order differential coefficients such as u, u', ϕ, ϕ', ϕ'' at a number of nodes along the beam as linear combinations of the highest derivatives in the differential equations of bending and torsion (u'' and $((EI_w)_t\phi'')'$ in eqns (2.14) and (2.15)). The minor-axis bending differential equation (eqn (2.14)) may be used to eliminate u'', in which case eqn (2.16) represents the torsion differential equation (eqn (2.15)) at the nodes, the unknowns being $\{((EI_w)_t\phi'')'\}$. The finite integral method was developed to obtain better accuracy than could be obtained by the finite difference method with the same nodes. This better accuracy arises principally because of the superior accuracy of numerical integration processes to those of numerical differentiation.

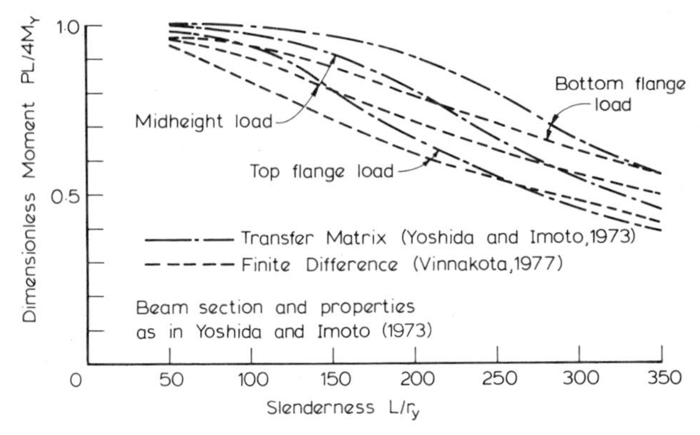

FIG. 2.13. Theoretical predictions of simple beam buckling moments.

The finite integral method has been compared with experimental results and the predictions of other theoretical methods for a very wide range of elastic buckling problems, and has been shown to be of high accuracy. It has been applied to a number of elastic and inelastic beam buckling problems by Kitipornchai and Trahair (1975*a*, *b*, *c*). Comparisons of its predictions for the inelastic buckling of hot-rolled simply supported beams with those of the finite difference method have already been shown in Fig. 2.12(a) and (b). Also shown in Fig. 2.12(b) are theoretical predictions obtained by the transfer matrix and finite element methods (Yoshida *et al.*, 1977). These are quite close to the finite integral predictions.

(*d*) *Transfer Matrix Method*

In the transfer matrix method (Tuma and Munshi, 1971), state vectors are used to represent the out-of-plane actions and deformations at the ends of the elements between nodes along the beam. These vectors are related by field and point transfer matrices. The field transfer matrix of an element is obtained from closed form solutions (Yoshida and Imoto, 1973) of the equations of out-of-plane bending and twisting of uniform elements under uniform in-plane moment, while the point transfer matrix represents the equilibrium and compatibility conditions at the node. The solution proceeds from the state vector at the left-hand end of the beam by alternately using field and point transfer matrices until the right-hand end is reached. The boundary conditions are then invoked, which reduces the equations to the general form of eqn (2.16), in which the vector $\{u\}$ represents the unknown out-of-plane actions and deformations at the beam ends.

The transfer matrix method has been used by Yoshida and Imoto (1973) to study the inelastic buckling of simply supported beams with various in-plane moment distributions, and of symmetric two-span continuous beams with uniformly distributed loads. Their predictions for hot-rolled simple beams with unbraced central concentrated loads are compared with those of Vinnakota (1977) in Fig. 2.13, and have been discussed previously in Section 2.3.2(b). Yoshida's predictions (Yoshida *et al.*, 1977) for a different set of hot-rolled simply supported beams with unbraced central concentrated loads are compared in Fig. 2.12(b) with other theoretical predictions and the experimental test results (Kitipornchai and Trahair, 1975*b*).

(e) Finite Element Method

In the finite element method for the lateral buckling of elastic beams (Barsoum and Gallagher, 1970; Rajasekaran, 1977), approximate fields for the out-of-plane deformations of an elemental length of a beam are used to derive the stiffness and stability matrices for the element. These are then combined using the nodal conditions of equilibrium and compatibility and then the boundary conditions are invoked. This leads to a set of out-of-plane stiffness equations of the form of eqn (2.18), in which the vector $\{u\}$ represents the unknown out-of-plane deformations at the nodes between the elements of the beam.

The application of the finite element method to the inelastic lateral buckling of steel beams has been studied extensively by Nethercot (1972*a*, *b*, 1973*a*, *b*, 1974*a*, *b*, 1975*a*, *b*, *c*). His theoretical predictions (Yoshida *et al.*, 1977) are compared in Fig. 2.12(b) with other theoretical predictions and the experimental test results (Kitipornchai and Trahair, 1975*b*) for hot-rolled simply supported beams with unbraced central concentrated loads (Fig. 2.11).

2.4 THEORETICAL PREDICTIONS OF INELASTIC BUCKLING

2.4.1 Effects of Residual Stresses

The effects of residual stresses on the inelastic lateral buckling of steel beams may be conveniently studied by considering simply supported beams in uniform in-plane bending (Fig. 2.1), firstly because in this case the modifying effects of moment gradient and non-uniform yielding are excluded, and secondly because the closed form solution

$$M_I = \sqrt{(\pi^2(EI_y)_t/L^2)}\{(\beta_x)_t\sqrt{(\pi^2(EI_y)_t)}/(2L) + \sqrt{[(GJ)_t + \pi^2(EI_w)_t/L^2 + \{(\beta_x)_t\sqrt{(\pi^2(EI_y)_t)}/2L\}^2]}\} \quad (2.20)$$

adapted from the elastic solution (Chapter 1) can be used.

Even so, the predictions of the inelastic buckling moment M_I vary not only with the magnitude and distribution of the residual stresses, but also with the beam cross-section, the yield stress, and the assumptions made when calculating the tangent modulus values of the various rigidities. These effects have been studied by many investigators, including Galambos (1963), Massey (1964*a*, *b*), Faella and Mazzolani (1972), Yoshida and Nishida (1972), Nethercot (1972*b*, 1974*a*), Djalaly (1974*a*) and Yoshida (1975).

The influences of variations of the cross-section and of the yield stress can be largely removed by constructing plots of the type shown in Fig. 2.14, in which the dimensionless moment at inelastic buckling M_I/M_Y is plotted

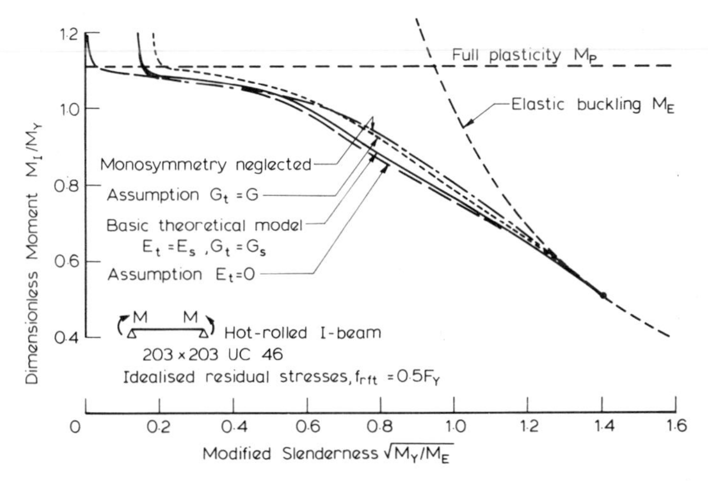

FIG. 2.14. Effects of assumptions made for inelastic buckling model.

against a modified slenderness $\sqrt{(M_Y/M_E)}$, in which M_E is the elastic buckling moment, and M_Y is the nominal moment at first yield when there are no residual stresses. This is certainly so in the elastic range, and approximately so in the inelastic range for statically determinate equal flange I-beams which have the same shape factor M_P/M_Y. (However, Nethercot (1973*b*) has demonstrated that the behaviour of mono-symmetric beams may be markedly different from that of equal flange beams.) The effect of variations in the distribution of the yield stress along the flanges has been shown to be small (Lindner and Bamm, 1977), while the effects of variations in the web yield stress are generally minimal, even in welded hybrid beams with low yield stress webs (Nethercot, 1975*c*).

The influences of the assumptions made concerning the inelastic rigidities are illustrated in Fig. 2.14 for a hot-rolled I-beam with idealised residual stresses (Trahair and Kitipornchai, 1972). The curve labelled 'basic theoretical model' was calculated using the assumptions discussed earlier in Section 2.2.3. It shows an almost linear increase in the inelastic buckling moment with decreasing slenderness, commencing at $M_I/M_Y = 0{\cdot}5$ when the compression flange tips first yield, to $M_I/M_Y \simeq 1{\cdot}08$ when both flanges are nearly fully yielded. For $M_I/M_Y > 1{\cdot}11$, the beam acts as if completely strain-hardened. The other curves of Fig. 2.14 show that the assumption of $G_t = G$ in the yielded regions leads to moderate increases in the buckling moment, and that the assumption that $E_t = 0$ in the yielded regions leads to slight decreases. More important are the overestimates of buckling moment caused by neglecting the effects of monosymmetry in the yielded cross-section.

The effects of the magnitude and the shape of the residual stress distribution are demonstrated in Fig. 2.15. The most important factor is the magnitude of the residual stress at the flange tips. High compressive residual stresses lead to early yielding at the compression flange tips, with significant reductions in the rigidities $(EI_y)_t$ and $(EI_w)_t$, and increases in the destabilising effect of monosymmetry caused by the lowering of the shear centre. Consequently, there are significant reductions in the inelastic buckling moments compared with those of annealed beams without residual stresses.

Yoshida (1977) has demonstrated that high tensile residual stresses at the flange tips of welded beams caused by flame-cutting lead to significant increases in the inelastic buckling strength. This is because the tensile residual stresses delay yielding of the compression flange tips, thereby maintaining the resistance at higher moments. On the other hand, early yielding at the tips of the tension flange has comparatively little effect

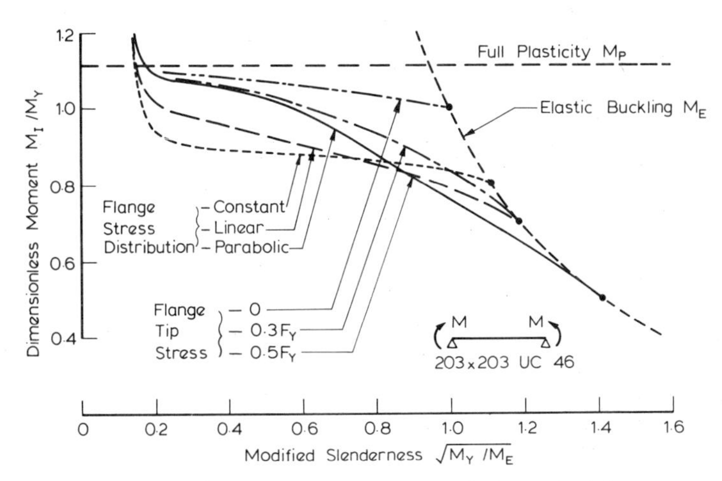

FIG. 2.15. Effects of residual stress on inelastic buckling.

because the tension flange deflects only slightly during buckling and makes only a small contribution to the buckling resistance. The effect of the tensile residual stresses at the flange tips is therefore to change the sense of monosymmetry, causing the shear centre to move upwards, and increasing the buckling resistance.

Also important in the effects of residual stresses is their distribution across the flanges, as the more nearly constant is the residual stress, then the more dramatic are the decreases in $(EI_y)_t$ and $(EI_w)_t$ once yielding commences (Yoshida, 1975, 1977). Thus, while the peak residual compressive stresses are often higher in hot-rolled beams than in welded beams (Fig. 2.3), causing earlier reductions in the buckling strength of intermediate slenderness beams, yet the more constant residual compressive stresses in welded beams cause greater reductions for low slenderness beams, as can be seen in Fig. 2.15.

2.4.2 Effects of Moment Distribution in Simple Beams

(a) General

Many theoretical studies have shown that the effect of variations of the major-axis moment distribution along a beam on its inelastic buckling resistance is very important. This is because yielding takes place only in the high moment regions, and so the consequent reductions in the section rigidities are localised. The inelastic buckling resistance depends markedly on the location and extent of these regions of reduced rigidity. The following subsections summarise some of these theoretical findings.

(b) End Moments

The inelastic buckling of beams with equal and opposite end moments has been discussed previously in Section 2.4.1. Studies of simply supported beams with unequal end moments M, βM include those by Massey and Pitman (1966), Massey (1967), Lay and Galambos (1967), Hartmann (1971), Yoshida and Nishida (1972), Yoshida and Imoto (1973), Djalaly (1974*a*, *b*), Nethercot (1975*a*), Yoshida (1975) and Dux and Kitipornchai (1978).

Nethercot and Trahair (1976*b*) used the inelastic buckling model developed in Section 2.2.3 and the finite element method (using elements with uniform properties) to study the inelastic buckling of several hot-rolled beams with idealised residual stresses (Fig. 2.16). The effect of the

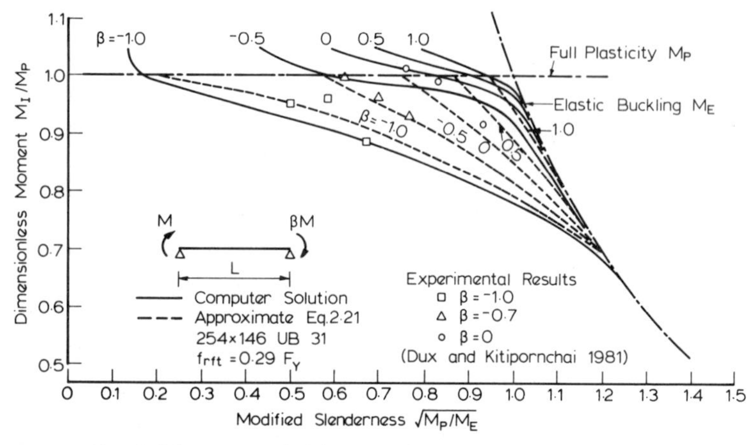

FIG. 2.16. Hot-rolled beams with unequal end moments.

end moment ratio β is very important, as the inelastic buckling moments of beams in uniform bending ($\beta = -1$) are reduced substantially below the elastic values, while those for beams with high moment gradient ($\beta \to 1$) are only slightly reduced, at least until the full plastic moment M_P is reached. This is because a high moment gradient ensures that yielding is confined to short regions at the ends of the beam. In this case, the central region, which is mostly responsible for the buckling resistance, remains elastic, and there is only a small reduction in buckling strength.

Also shown in Fig. 2.16 are simple approximations given by

$$\frac{M_I}{M_P} \simeq 0{\cdot}70 + \frac{0{\cdot}30(1 - 0{\cdot}70 M_P/M_E)}{(0{\cdot}61 - 0{\cdot}30\beta + 0{\cdot}07\beta^2)} \tag{2.21}$$

in which M_E is the elastic buckling moment for the same value of β (Chapter 1). These approximations, which are valid in the range $0{\cdot}7 \leq M_I/M_P \leq 1{\cdot}0$, are generally conservative, except for the small discrepancies which occur for uniform bending ($\beta = -1$). They can be used to determine limiting modified slendernesses at which the inelastic buckling moment M_I is equal to the full plastic moment M_P. These limiting values are

$$\sqrt{(M_P/M_E)_P} = \sqrt{\{(0{\cdot}39 + 0{\cdot}30\beta - 0{\cdot}07\beta^2)/0{\cdot}70\}} \qquad (2.22)$$

and vary from 0·17 to 0·94 as β varies from -1 to $+1$.

The experimental results obtained by Dux and Kitipornchai (1981) for beams with central concentrated braced loads (which are equivalent to half-length beams with $\beta = 0$) are also shown in Fig. 2.16. It can be seen that these are reasonably close to the theoretical predictions.

(c) Transverse Loads

Studies of the inelastic buckling of simply supported beams with transverse concentrated or distributed loads which are free to move laterally with their beams include those by Massey and Pitman (1966), Massey (1967), Hartmann (1971), Faella and Mazzolani (1972), Yoshida and Imoto (1973), Djalaly (1974*a*, *b*), Nethercot (1975*a*), Kitipornchai and Trahair (1975*a*) and Yoshida (1975, 1977). Studies have also been made of the effects of eccentric and lateral loads on the inelastic biaxial bending of beams (Lindner, 1971, 1974).

Nethercot and Trahair (1976*b*) studied the inelastic buckling of simply supported beams with loads acting at the geometrical axis that were either central, or symmetrically placed (a from each end), or uniformly distributed. They compared their predictions for beams with central concentrated load with the approximate solutions of eqn (2.21) for beams in uniform bending (Fig. 2.17), and found that the former were a little higher. In all cases, yielding takes place in the mid-span region, where it causes the greatest reductions in buckling resistance. They proposed that these loading situations could be treated as equivalent end moments M and βM by using $\beta = -0{\cdot}70$ for central concentrated load, $\beta = -0{\cdot}90$ for uniformly distributed load, and

$$\beta = -1{\cdot}00 + 0{\cdot}60a/L \qquad (2.23)$$

for beams with two symmetrical concentrated loads at distances a from the ends. This would then permit the approximate maximum moments M_I at inelastic buckling to be calculated from eqn (2.21).

These studies by Nethercot and Trahair (1976*b*) also suggested that these

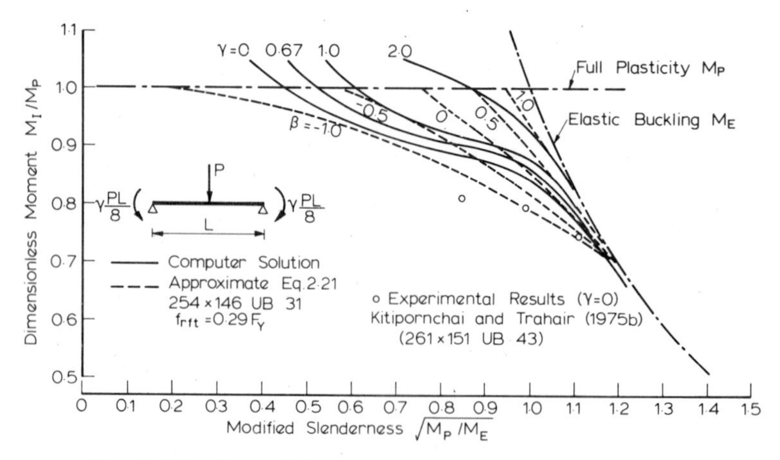

FIG. 2.17. Hot-rolled beams with unbraced central loads.

approximations could be used for beams with loads acting away from their geometrical axes, provided that appropriate modifications (Chapter 1) were made when calculating the elastic buckling moments M_E to be used in eqn (2.21). Some experimental evidence on this is shown by the results of Kitipornchai and Trahair (1975*b*) plotted in Fig. 2.17.

(*d*) *End Moments and Transverse Loads*

The inelastic buckling of simply supported beams with unbraced transverse loads and end moments has been studied by a number of investigators, including Yoshida and Imoto (1973), Yoshida (1977) and Nakamura and Wakabayashi (1981). The results of these studies may be used to approximate the inelastic buckling loads of continuous beams in which major-axis continuity between adjacent spans induces major-axis end restraining moments, when the effects of continuity on minor-axis bending and warping are neglected.

Nethercot and Trahair (1976*b*) studied the effects of equal restraining end moments $\gamma PL/8$ on the elastic buckling of hot-rolled beams with central concentrated loads (Fig. 2.17). They found that for high end moments ($\gamma \rightarrow 2{\cdot}0$) the behaviour was similar to that of beams with high moment gradient ($\beta \rightarrow 0{\cdot}5$), since yielding occurred only at the ends where it was relatively unimportant. This led them to propose that these loading arrangements could be regarded as equivalent unequal end moments, with

$$\beta = -0{\cdot}70 + 0{\cdot}30\gamma^2 \tag{2.24}$$

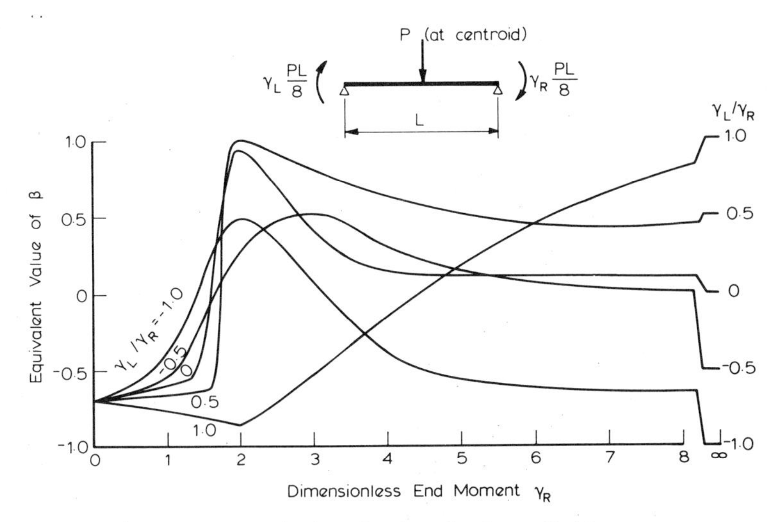

FIG. 2.18. Equivalent β values for hot-rolled beams.

In another study (Nethercot and Trahair, 1977), the effects of unequal end moments $\gamma_L PL/8$ and $\gamma_R PL/8$ on the inelastic buckling of hot-rolled beams with central concentrated loads P were investigated. The results of this study are summarised by the equivalent end moment ratios β shown in Fig. 2.18 for different values of γ_R and γ_L/γ_R.

2.4.3 Cantilevers

The inelastic buckling of hot-rolled cantilevers with concentrated end loads or uniformly distributed loads has been investigated by Nethercot (1975*a*). This study included the effect of the height of the loading relative to the geometrical axis of the cantilever. In all cases, yielding was confined to the support region, and caused substantial reductions from the elastic buckling resistance which were similar to those for simply supported beams with central concentrated loads.

2.4.4 Effects of Continuity in Determinate Braced Beams

In a statically determinate braced beam, there may be an interaction between adjacent segments during inelastic buckling which is similar to that which occurs in an elastic beam (Chapter 1). Nethercot and Trahair (1976*b*, 1977) developed an extension of their elastic method (Nethercot and Trahair, 1976*a*; Trahair, 1977) for approximating the buckling resistance. The accuracy of this extension is demonstrated in Fig. 2.19.

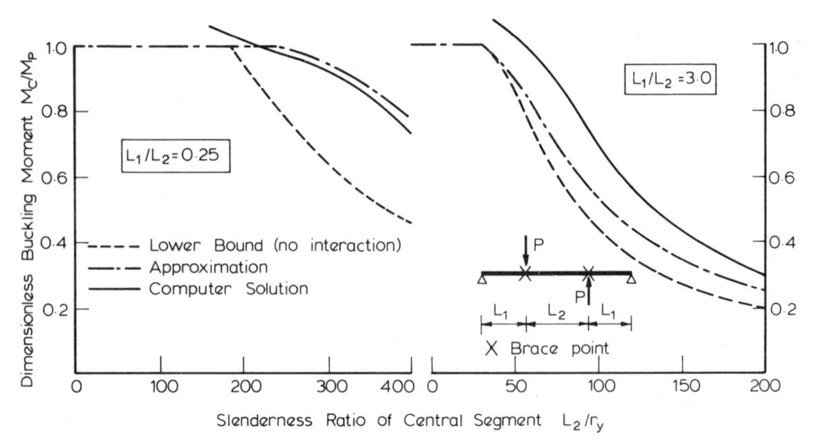

FIG. 2.19. Inelastic buckling of braced beams.

Dux and Kitipornchai (1981) carried out experiments on the inelastic buckling of braced beams, and their results are shown in Fig. 2.16. These are in reasonable agreement with the theoretical predictions of Nethercot and Trahair (1976*b*), despite the opposing effects of geometrical imperfections and minor-axis and warping continuity which are omitted from the theory. Theoretical and experimental studies were also made by Nakamura and Wakabayashi (1981).

2.4.5 Continuous Beams

A continuous beam is statically indeterminate, and as yielding progresses in the beam, its major-axis bending moments are redistributed from the elastic distribution towards that of the plastic collapse mechanism. (However, this redistribution may not be as rapid as is suggested by simple plastic theory because of the effects of strain-hardening.) The redistribution allows the loads to increase to levels which are often significantly higher than the nominal loads which cause first yield in beams without residual stresses. Because of this, there are usually several areas of the beam with appreciable yielding. The changes in both the moment and yield region distributions may be favourable or unfavourable with respect to the inelastic buckling resistance.

Theoretical predictions of the inelastic buckling loads of two-span continuous hot-rolled beams are shown in Fig. 2.20 (Yoshida *et al.*, 1977), together with the corresponding predictions for simply supported beams with central concentrated loads. All of the plotted points (with the

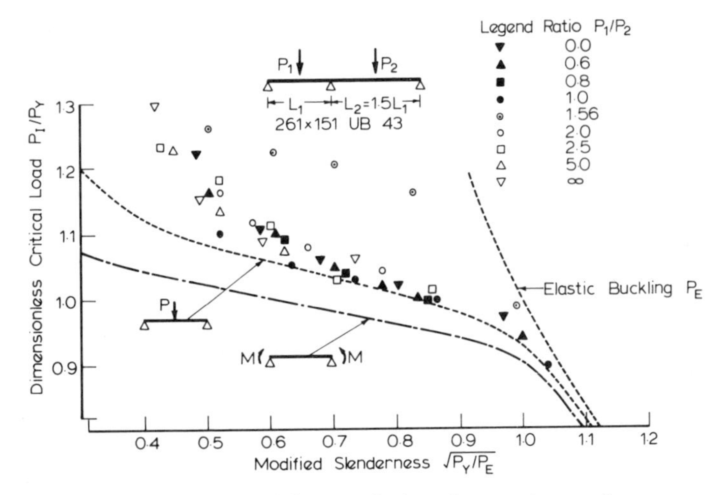

FIG. 2.20. Inelastic buckling predictions for continuous beams.

exception of those for the load ratio $P_1/P_2 = 1{\cdot}56$) are very closely grouped, and for modified slendernesses $\sqrt{(P_Y/P_E)} > 0{\cdot}6$ are quite close to the curve for simply supported beams. This is because all of these continuous beams yield first at one of the load points at mid-span. For decreasing slenderness, the plotted points rise more rapidly than the curve for simply supported beams, principally because yielding spreads more slowly from the load points as a result of the higher moment gradients, particularly towards the interior support.

For continuous beams with load ratios of $P_1/P_2 = 1{\cdot}56$, yielding occurs first at the interior support, where it causes comparatively small reductions from the elastic buckling resistance, in much the same way as do the end moments of simple beams with high moment gradient (Fig. 2.16). Thus the inelastic buckling resistances of these beams are noticeably higher than those of the other continuous beams shown in Fig. 2.20.

Some of these theoretical predictions for two-span continuous beams (for $L_1 = 2{\cdot}44\,\text{m}$) are also shown in Fig. 2.21, together with the experimental results for beams tested by Poowannachaikul and Trahair (1976). The experimental results are surprisingly low, and even more so than those shown in Fig. 2.17 for simple beams of the same cross-section (Kitipornchai and Trahair, 1975*b*). No satisfactory explanation has yet been advanced for these discrepancies.

Approximate theoretical predictions for the inelastic buckling of continuous beams with central concentrated loads may be made by using

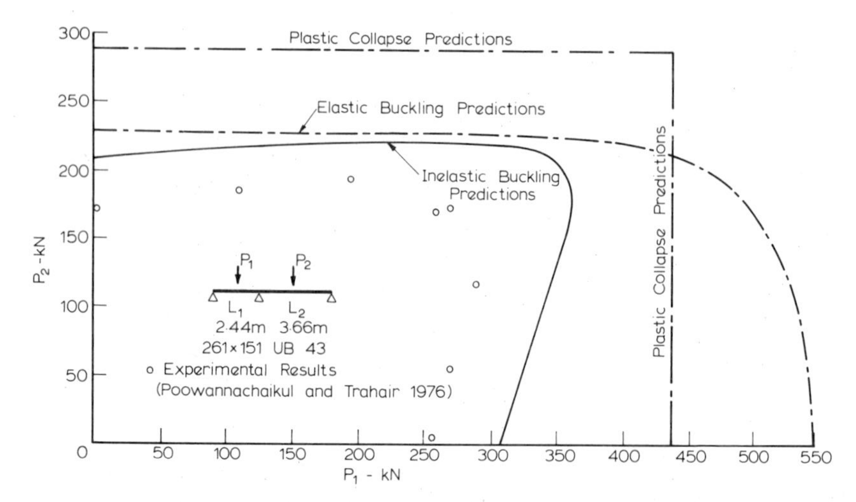

FIG. 2.21. Experimental maximum loads for continuous beams.

the method of Nethercot and Trahair (1976*b*, 1977) and the information given in Fig. 2.18, together with eqn (2.21), provided that the major-axis moment distribution can be determined. It is suggested that two approximations should be obtained, one for the elastic moment distribution, and one for the distribution at plastic collapse. If necessary, the final approximation may be determined by interpolation.

REFERENCES

ABDEL-SAYED, G. and AGLAN, A. A. (1973) Inelastic lateral torsional buckling of beam columns. *Publications, IABSE*, **33**(II), 1–16.

BARSOUM, R. S. and GALLAGHER, R. H. (1970) Finite element analysis of torsional and torsional-flexural stability problems. *International Journal for Numerical Methods in Engineering*, **2**, 335–52.

BISHOP, R. E. D., GLADWELL, G. M. L. and MICHAELSON, S. (1965) *The Matrix Analysis of Vibration*, Cambridge University Press, Cambridge.

BOOKER, J. R. and KITIPORNCHAI, S. (1971) Torsion of multi-layered rectangular sections. *Journal of the Engineering Mechanics Division, ASCE*, **97**(EM5), Proc. Paper 8415, 1451–68.

BRADFORD, M. A. and TRAHAIR, N. S. (1981) Distortional buckling of I-beams. *Journal of the Structural Division, ASCE*, **107**(ST2), Proc. Paper 16057, 355–70.

BROWN, P. T. and TRAHAIR, N. S. (1968) Finite integral solution of differential equations. *Civil Engineering Transactions, Institution of Engineers, Australia*, **CE10**(2), 193–6.

Collatz, L. (1966) *Functional Analysis and Numerical Mathematics*, Academic Press, New York.

Djalaly, H. (1974*a*) Calcul de la résistance ultime au déversement. *Construction Métallique*, No. 1, 58–77.

Djalaly, H. (1974*b*) Calcul de la résistance ultime au déversement. *Construction Métallique*, No. 4, 54–61.

Dux, P. F. and Kitipornchai, S. (1978) Approximate inelastic buckling moments for determinate I-beams. *Civil Engineering Transactions, Institution of Engineers, Australia*, **CE20**(2), 128–33.

Dux, P. F. and Kitipornchai, S. (1981) Inelastic beam buckling experiments. Research Report CE24, Dept of Civil Engineering, University of Queensland, Australia.

Dwight, J. B. and White, J. D. (1977) Prediction of weld shrinkage stresses in plated structures. *Preliminary Report, 2nd International Colloquium on Stability of Steel Structures*, ECCS–IABSE, Liège, pp. 31–7.

Faella, C. and Mazzolani, F. M. (1972) Instabilita flesso-torsionale inelastica di travi metalliche sotto carichi trasversali. *Construzioni Metalliche*, pp. 1–18.

Fukumoto, Y. and Galambos, T. V. (1966) Inelastic lateral-torsional buckling of beam-columns. *Journal of the Structural Division, ASCE*, **92**(ST2), Proc. Paper 4770, 41–61.

Fukumoto, Y. and Itoh, Y. (1981) Statistical study of experiments on welded beams. *Journal of the Structural Division, ASCE*, **107**(ST1), Proc. Paper 15965, 89–103.

Fukumoto, Y., Itoh, Y. and Kubo, M. (1980) Strength variation of laterally unsupported beams. *Journal of the Structural Division, ASCE*, **106**(ST1), Proc. Paper 15142, 165–81.

Galambos, T. V. (1963) Inelastic lateral buckling of beams. *Journal of the Structural Division, ASCE*, **89**(ST5), Proc. Paper 3683, 217–42.

Hancock, G. J. (1978) Local, distortional, and lateral buckling of I-beams. *Journal of the Structural Division, ASCE*, **100**(ST11), Proc. Paper 14155, 2205–22.

Hartmann, A. J. (1971) Inelastic flexural–torsional buckling. *Journal of the Engineering Mechanics Division, ASCE*, **97**(EM4), Proc. Paper 8294, 1103–19.

Jarnot, F. T. and Young, B. W. (1977) The ultimate load capacity of laterally unsupported beams subjected to equal and unequal terminal moments. *Preliminary Report, 2nd International Colloquium on Stability of Steel Structures*, ECCS–IABSE, Liège, pp. 217–22.

Johnson, C. P. and Will, K. M. (1974) Beam buckling by finite element procedure. *Journal of the Structural Division, ASCE*, **100**(ST3), Proc. Paper 10432, 669–85.

Kitipornchai, S. and Trahair, N. S. (1975*a*) Buckling of inelastic I-beams under moment gradient. *Journal of the Structural Division, ASCE*, **101**(ST5), Proc. Paper 11295, 991–1004.

Kitipornchai, S. and Trahair, N. S. (1975*b*) Inelastic buckling of simply supported steel I-beams. *Journal of the Structural Division, ASCE*, **101**(ST7), Proc. Paper 11419, 1333–47.

Kitipornchai, S. and Trahair, N. S. (1975*c*) Elastic behaviour of tapered monosymmetric I-beams. *Journal of the Structural Division, ASCE*, **101**(ST8), Paper 11479, 1661–78.

LAY, M. G. (1965) Flange local buckling in wide-flange shapes. *Journal of the Structural Division, ASCE*, **91**(ST6), Proc. Paper 4554, 95–116.

LAY, M. G. and GALAMBOS, T. V. (1967) Inelastic beams under moment gradient. *Journal of the Structural Division, ASCE*, **93**(ST1), Proc. Paper 5110, 381–99.

LAY, M. G. and WARD, R. (1969) Residual stresses in steel structures. *Steel Construction*, **3**(3), 2–21.

LEE, G. C., FINE, D. S. and HASTREITER, W. R. (1967) Inelastic torsional buckling of H-columns. *Journal of the Structural Division, ASCE*, **93**(ST5), Proc. Paper 5517, 295–307.

LINDNER, J. (1971) Näherungsweise Ermittlung der Traglasten von auf Biegung und Torsion beanspruchten I-Trägern. *Die Bautechnik*, **48**(5), 160–70.

LINDNER, J. (1974) Traglastkurven für I-Träger, die durch aussermittige Querlasten beansprucht werden. *Der Stahlbau*, **43**(10), 307–13.

LINDNER, J. and BAMM, D. (1977) Influence of realistic yield stress distribution on lateral torsional buckling loads. *Preliminary Report, 2nd International Colloquium on Stability of Steel Structures*, ECCS–IABSE, Liège, pp. 213–16.

MASSEY, P. C. (1964*a*) The lateral stability of steel I-beams in the plastic range. *Civil Engineering Transactions, Institution of Engineers, Australia*, **CE6**(2), 119–29.

MASSEY, P. C. (1964*b*) The effect of residual stress on the lateral stability of steel I-beams. *New Zealand Engineering*, **19**(9), 338–42.

MASSEY, P. C. (1967) Inelastic lateral stability of open section steel beams under non-uniform bending moments. *New Zealand Engineering*, **22**(4), 135–40.

MASSEY, P. C. and PITMAN, F. S. (1966) Inelastic lateral stability under a moment gradient. *Journal of the Engineering Mechanics Division, ASCE*, **92**(EM2), Proc. Paper 4779, 101–11.

NAKAMURA, T. and WAKABAYASHI, M. (1981) Lateral buckling of beams braced by purlins. *Proceedings, US–Japan Seminar on Inelastic Stability of Steel Structures and Structural Elements*, Tokyo.

NETHERCOT, D. A. (1972*a*) Recent progress in the application of the finite element method to problems of the lateral buckling of beams. *Proceedings, Conference on Finite Element Methods in Civil Engineering*, Engineering Institute of Canada, Montreal (June), p. 367.

NETHERCOT, D. A. (1972*b*) Factors affecting the buckling stability of partially plastic beams. *Proceedings, Institution of Civil Engineers*, **53**, 285–304.

NETHERCOT, D. A. (1973*a*) The solution of inelastic lateral stability problems by the finite element method. *Proceedings, 4th Australasian Conference on Mechanics of Structures and Materials*, Brisbane, pp. 183–90.

NETHERCOT, D. A. (1973*b*) Inelastic buckling of monosymmetric I-beams. *Journal of the Structural Division, ASCE*, **99**(ST7), Technical Note, 1696–1701.

NETHERCOT, D. A. (1974*a*) Residual stresses and their influence upon the lateral buckling of rolled steel beams. *Structural Engineer*, **52**(3), 89–96.

NETHERCOT, D. A. (1974*b*) Buckling of welded beams and girders. *Publications, IABSE*, **34**(I), 107–21.

NETHERCOT, D. A. (1975*a*) Inelastic buckling of steel beams under non-uniform moment. *Structural Engineer*, **53**(2), 73–8.

NETHERCOT, D. A. (1975*b*) Effective lengths of partially plastic steel beams. *Journal of the Structural Division, ASCE*, **101**(ST5), Technical Note, 1163–6.

NETHERCOT, D. A. (1975*c*) Bending and buckling of welded hybrid beams. Research Report R41, Dept of Civil and Structural Engineering, University of Sheffield (Mar.).

NETHERCOT, D. A. and TRAHAIR, N. S. (1976*a*) Lateral buckling approximations for elastic beams. *Structural Engineer*, **54**(6), 197–204.

NETHERCOT, D. A. and TRAHAIR, N. S. (1976*b*) Inelastic lateral buckling of determinate beams. *Journal of the Structural Division, ASCE*, **102**(ST4), Proc. Paper 12020, 701–17.

NETHERCOT, D. A. and TRAHAIR, N. S. (1977) Lateral buckling calculations for braced beams. *Civil Engineering Transactions, Institution of Engineers, Australia*, **CE19**(2), 211–14.

POOWANNACHAIKUL, T. and TRAHAIR, N. S. (1976) Inelastic buckling of continuous steel I-beams. *Civil Engineering Transactions, Institution of Engineers, Australia*, **CE18**(2), 134–9.

RAJASEKARAN, S. (1977) Finite element method for plastic beam-columns. Chap. 12 in *Theory of Beam-Columns*, Vol. 2: *Space Behaviour and Design*, ed. W.-F. Chen and T. Atsuta, McGraw-Hill, New York, pp. 539–608.

SHANLEY, F. R. (1947) Inelastic column theory. *Journal of Aeronautical Sciences*, **14**(5), 261–8.

TRAHAIR, N. S. (1977) *Behaviour and Design of Steel Structures*, Chapman and Hall, London.

TRAHAIR, N. S. and KITIPORNCHAI, S. (1972) Buckling of inelastic I-beams under uniform moment. *Journal of the Structural Division, ASCE*, **98**(ST11), Proc. Paper 9339, 2551–66.

TUMA, J. J. and MUNSHI, R. K. (1971) *Theory and Problems of Advanced Structural Analysis*, McGraw-Hill, New York.

VINNAKOTA, S. (1977) Finite difference method for plastic beam-columns. Chap. 10 in *Theory of Beam-Columns*, Vol. 2: *Space Behaviour and Design*, ed. W.-F. Chen and T. Atsuta, McGraw-Hill, New York, pp. 451–503.

WHITE, M. W. (1960) Inelastic lateral instability of beams and their bracing requirements. Thesis, presented to Lehigh University, Bethlehem, Pa.

WILKINSON, J. H. (1960) Householder's method for the solution of the algebraic eigenproblem. *Computer Journal*, **3**, 23–7.

YOSHIDA, H. (1975) Lateral buckling strength of plate girders. *Publications, IABSE*, **34**(I), 107–20.

YOSHIDA, H. (1977) Buckling curves for welded beams. *Preliminary Report, 2nd International Colloquium on Stability of Steel Structures*, ECCS–IABSE, Liège, pp. 191–6.

YOSHIDA, H. and IMOTO, Y. (1973) Inelastic lateral buckling of restrained beams. *Journal of the Engineering Mechanics Division, ASCE*, **99**(EM2), Proc. Paper 9666, 343–66.

YOSHIDA, H. and NISHIDA, S. (1972) The effect of residual stresses on the lateral-torsional buckling of steel beam-columns. Research Report SM11, Dept of Civil Engineering, Kanazawa University, Japan.

YOSHIDA, H., NETHERCOT, D. A. and TRAHAIR, N. S. (1977) Analysis of inelastic buckling of continuous beams. *Proceedings, IABSE*, No. P-3/77, 1–14.

YOUNG, B. W. (1975) Residual stresses in hot-rolled sections. *Proceedings, International Colloquium on Column Strength, IABSE*, **23**, 25–38.

Chapter 3

DESIGN OF LATERALLY UNSUPPORTED BEAMS

D. A. NETHERCOT

Department of Civil and Structural Engineering,
University of Sheffield, UK

and

N. S. TRAHAIR

School of Civil and Mining Engineering, University of Sydney,
New South Wales, Australia

SUMMARY

Available experimental data on the elastic and inelastic lateral buckling of steel beams is reviewed and its use in the preparation of design approaches discussed. Several representative design methods for laterally unsupported beams are summarised and new developments in this area, in particular the trend towards the use of a limit state format, are explained.

NOTATION

h	Distance between flange centroids
l	Effective length
m	Moment modification factor
n	Index in eqn (3.10)
r_T	Radius of gyration of compression flange plus one-sixth of the web
r_y	Minor-axis radius of gyration

t	Web thickness
u	Buckling parameter (eqn (3.23))
v	See eqn (3.23)
x	Torsional index (eqn (3.23))
A	Cross-sectional area
B	Flange width
D	Depth of section
E	Young's modulus of elasticity
F_{bc}	Maximum permissible compressive bending stress
F_L	Nominal stress at first yield
F_Y	Yield stress
F_{0b}	Elastic buckling stress $= M_E/Z_x$
G	Shear modulus of elasticity
I_w	Warping section constant
I_x, I_y	Second moments of area about the x, y axes
J	Torsion section constant
M	Moment
M_c	Buckling moment
M_E	Elastic buckling moment
M_I	Inelastic buckling moment
M_P	Full plastic moment
M_r	First yield moment for beam with residual stress (eqn (3.27))
M_u	Moment at failure
M_{u0}	Value of M_u for beams in uniform bending
M_Y	Nominal first yield moment $= F_Y Z_x$
M_0	Elastic buckling moment of simple beam in uniform bending
P_E	Elastic buckling load
P_P	Load at plastic collapse
Q	Load effect
Q_k	Nominal load effects
R	Resistance
R_n	Nominal resistance
T	Flange thickness
Z_p	Major-axis plastic section modulus
Z_x	Major-axis elastic section modulus
β	Ratio of end moments, or safety index
γ	Coefficient defining value of end moment
γ_k	Load factor

γ_L, γ_R Values of γ for left and right ends
η Imperfection parameter (eqns 3.12, 3.14, 3.21)
λ Geometrical slenderness ratio $= l/r_y$
λ_{LT} Lateral slenderness ratio (eqn (3.22))
λ_P See eqn (3.28)
λ_r Value of λ when $M_0 = M_r$
ϕ Resistance factor

3.1 INTRODUCTION

Satisfactory design of laterally unsupported beams requires a proper consideration of all those factors which influence their capacity to carry load as explained in Chapters 1 and 2. This must include some aspects of the problem which cannot, at present, be fully allowed for theoretically, but which are known to be important. Thus it is known that real beams contain residual stresses; are not initially perfectly straight; are not geometrically perfect in the sense that the web may be offset, the flanges may be of unequal size, etc. (making a nominally symmetrical section into an unsymmetrical one); have material properties that change with location within the cross-section; and possess various other forms of 'imperfection'. However, it is not possible to account for all of these items accurately in a theoretical solution. Moreover, all of these quantities are not deterministic, but exhibit a certain degree of scatter, such as the variability of the residual stress produced by small differences in cooling conditions after hot rolling, which is beyond the control of the designer. Thus even if a very powerful and precise analytical approach were available, it could only be expected to predict accurately the lateral buckling strength corresponding exactly to the data used in the analysis. Before it could be used as the basis for a design approach, several analyses, covering the range of values of each significant variable entering the problem, would be necessary. Whilst such probabilistic studies have been conducted for the much simpler problem of the flexural buckling of steel columns (Bjhovde, 1978), their application to beam lateral buckling does not appear feasible at present, on account of the lack of a wholly adequate theoretical model of the important effects, of a lack of sufficiently comprehensive basic data on member properties—especially their variability—and of the excessive demands that such a study would make on currently available computing facilities.

In seeking to devise design approaches, it is natural therefore to turn to test data as at least part of the basis. Indeed, some quite satisfactory design procedures owe very little directly to theoretical studies, being based almost entirely on experimental results. The danger of such methods is, of course, that their region of application is effectively limited to the range of parameters covered by the available data. Even within this region, there is a danger that lack of data in certain areas, or incorrectly evaluated data, or the use of test results for which the experimental conditions did not properly conform to those of the application, will make this wholly empirical method a somewhat unreliable approach.

The best design approaches therefore employ both theoretical and experimental studies, together with judgements of their limitations. In addition, an appreciation is needed of the requirements of the designer and the standards of construction that may reasonably be expected. The objective is therefore to produce a method that combines technical accuracy with ease of application and realistic fabrication demands. Several such methods are discussed in Section 3.3 of this chapter. This discussion is preceded in Section 3.2 by a review of some of the available test data relating to lateral buckling strength. This review supplements the material on theoretical studies of elastic and inelastic lateral buckling presented in Chapters 1 and 2.

3.2 REVIEW OF EXPERIMENTAL DATA

3.2.1 General

Experimental studies of the buckling of steel beams are generally of two types. The first of these is intended to provide correlations with theoretical analyses of the buckling of beams which are geometrically perfect, and to demonstrate such influences as the pattern of moments and the type of support conditions. In the case of inelastic buckling, additional factors such as residual stresses may also be investigated. For these experiments, the geometrical imperfections should be very small, and countered where possible by the eccentricities of the applied loads. Some of these experiments have already been referred to in the previous chapters and their results are shown in Figs. 2.12, 2.16, 2.18 and 2.21 of Chapter 2.

The second type of experimental study consists of tests conducted to determine the maximum load-carrying capacity of real beams. These beams should have representative geometrical imperfections and load eccentricities, so that they behave as shown in Fig. 3.1, with lateral deflections and twists increasing with load, slowly at first, and then more rapidly as the maximum load is approached (see eqn (1.27) and Fig. 1.14 of

Chapter 1). This maximum load is less than the buckling load by an amount which depends on the magnitude of the imperfections, eccentricities and residual stresses. These tests furnish valuable information not only for developing design strength formulations, but also for assessing the accuracy of analyses of the inelastic biaxial bending of beams. Fukumoto (1981) has proposed that a numerical data bank be established in which the results of these tests can be kept. In the following subsections, the results of early and recent tests are discussed.

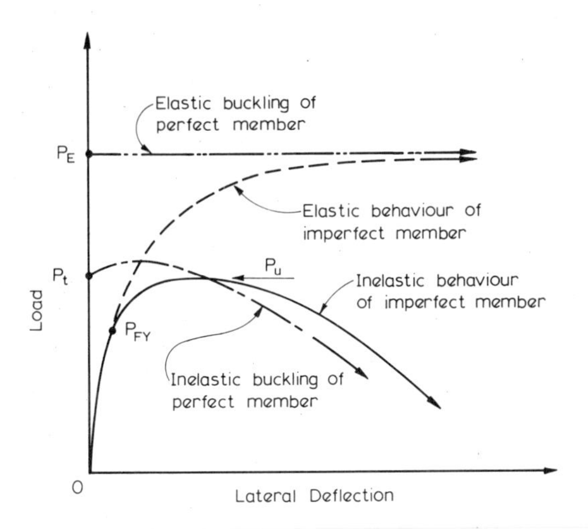

FIG. 3.1. Behaviour of real beams.

3.2.2 Slender Beams: Elastic Buckling

Slender beams are defined as those for which buckling occurs whilst the material is still elastic. The theories of Chapter 1 are therefore appropriate. Evaluations of several series of tests on slender beams have been produced by Galambos and Ravindra (1974) and Yura *et al.* (1978). These cover a number of sources and are mainly of the type which seek to provide correlation with theoretical buckling analyses. A summary of their findings in the form of a histogram of the ratios of test load to predicted load is shown in Fig. 3.2. For the 185 tests considered, the mean m is 1·03 and the standard deviation s is 0·10. These values show good correlation between elastic buckling theory and laboratory tests.

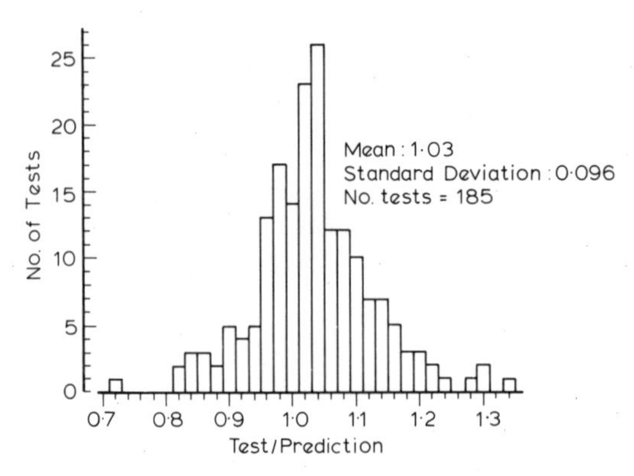

FIG. 3.2. Correlation between elastic buckling theory and laboratory test results (Galambos and Ravindra, 1974).

3.2.3 Stocky Beams: Plastic Collapse

Provided the individual plate elements of the cross-section are proportioned so as to avoid local buckling, beams of low slenderness will fail by in-plane plastic collapse. In this case, lateral buckling considerations are only necessary to define the maximum slenderness of beams which can reach plastic collapse. In view of this, the relevant experimental data will not be reviewed here. Instead, these maximum slendernesses for plastic beams will be considered as the lower slendernesses for beams which fail by inelastic lateral buckling. These are discussed in Section 3.2.4.

A related, though not identical, problem is the determination of the maximum slenderness for which the rotation capacity required for the redistribution of moments assumed in plastic design is also available (Ojalvo and Weaver, 1978).

3.2.4 Beams of Intermediate Slenderness: Inelastic Buckling

(*a*) *Early Tests*

Although a few series of tests have been conducted specifically to check the tangent modulus approach to the inelastic buckling of initially straight beams using specially prepared specimens, the great majority have used normal, commercially available sections. Many of these experimental studies of the buckling of commercial beams have been reviewed by Fukumoto and Kubo (1977*a*, *b*, *c*). Their review includes 95 tests on hot-rolled beams and 89 on welded beams and girders which they classified as being in the inelastic range.

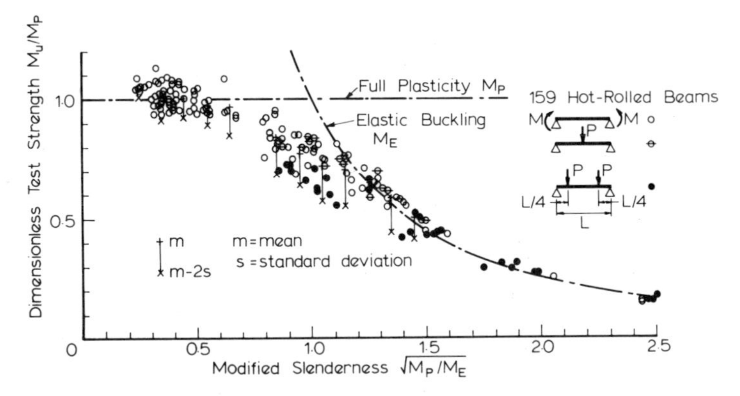

FIG. 3.3. Test results for hot-rolled beams.

The results of these tests are shown in Figs. 3.3 and 3.4. In these figures, variations in the elastic range due to beam geometry, support and loading conditions, and restraints are eliminated by plotting the non-dimensional maximum moment M_u/M_P against the modified slenderness $\sqrt{(M_P/M_E)}$. Nevertheless, there is considerable scatter in each figure, demonstrating not only the effects of imperfections and load eccentricities, but also those of residual stress, moment distribution and lateral continuity between adjacent segments on inelastic beams.

Points representing the mean (m) and the mean minus two standard deviations ($m - 2s$) of these tests are shown in Fig. 3.5. It can be seen that the mean strengths of the hot-rolled beams are higher than those of the welded beams, and that their standard deviations are generally lower.

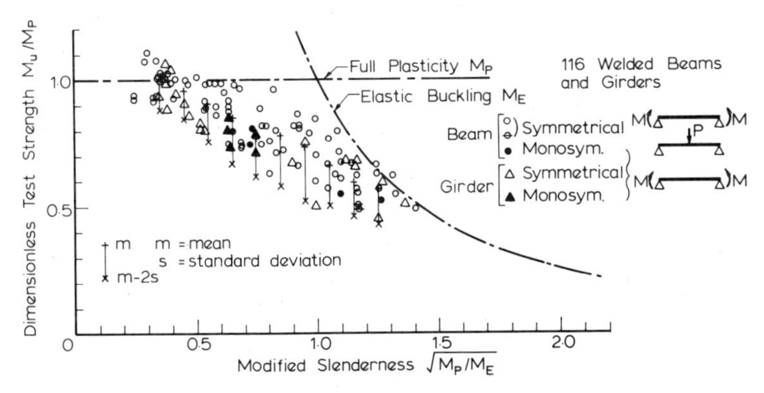

FIG. 3.4. Test results for welded beams and girders.

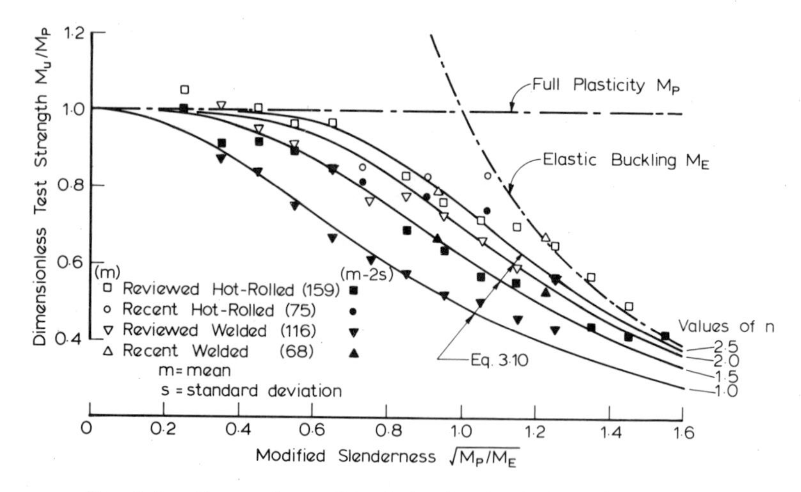

FIG. 3.5. Comparison of statistics for determinate beam tests.

Both Fig. 3.3 and Fig. 3.4 illustrate the existence of a plateau region at low slenderness within which the beam's full in-plane strength is attained as discussed in the previous section.

(*b*) *Recent Tests*

(*i*) *Statically determinate beams.* Fukumoto *et al.* (1980) tested 75 hot-rolled steel beams, all having nominally identical cross-section dimensions, material properties and residual stresses. Three different spans were used, and each beam was tested with an unbraced concentrated load acting at the centre of the top flange. Extensive measurements were made of the dimensions, material properties, residual stresses and initial crookednesses and twists. However, no information was given on load eccentricities or overall twists. Subsequently, Fukumoto and Itoh (1981) made similar tests on 68 welded steel beams, using two different spans.

In general, these beams had low geometrical imperfections (mean crookednesses of 0·084 mm m^{-1} and 0·125 mm m^{-1} for the hot-rolled and welded beams respectively), and low flange residual stresses (means of $0{\cdot}08F_Y$ at the tips of the hot-rolled beams, and $0{\cdot}121F_Y$ over the outer portions of the welded beams). The values of the mean (m) and mean minus two standard deviations ($m - 2s$) of the test strengths are compared in Fig. 3.5 with those of the reviewed tests on hot-rolled beams and welded beams and girders (Fukumoto and Kubo, 1977*a*, *b*, *c*). It can be seen that the values for the recent tests are generally higher than those of the reviewed tests. It can also be seen that the standard deviations of the recent tests are smaller than those of the reviewed tests, reflecting the fewer variables and tighter control of the recent tests.

Very recently, Wakabayashi and Nakamura (1983) reported a series of 18 tests on pairs of welded beams, simply supported at two points, and cantilevered beyond these supports. Each pair was laterally braced at the ends and the supports. Four pairs were also braced at mid-span, while five other pairs were braced by purlins. The test results demonstrate the importance of moment distribution and bracing on the strength of beams.

(*ii*) *Continuous beams.* A series of 21 two-span continuous welded beams was tested by Fukumoto *et al.* (1982), who again made extensive measurements of the dimensions, material properties, residual stresses and initial crookednesses and twists. Their test results are compared in Fig. 3.6 with those on hot-rolled continuous beams tested by Poowannachaikul and Trahair (1976). The low values of these latter test results have already been discussed in Section 2.4.5 of Chapter 2.

It can be seen that most of the new test results are for comparatively slender beams, and therefore give comparatively little information on the inelastic buckling behaviour. Also shown in Fig. 3.6 are two theoretical curves (eqn (2.2) of Chapter 2) for simply supported hot-rolled beams with end moments M and βM. An examination of the bending moment distributions of the more critical spans of the continuous beams suggests that they fall into the range $0 < \gamma_R < 1{\cdot}5$ with $\gamma_L = 0$, as defined in Fig. 2.18

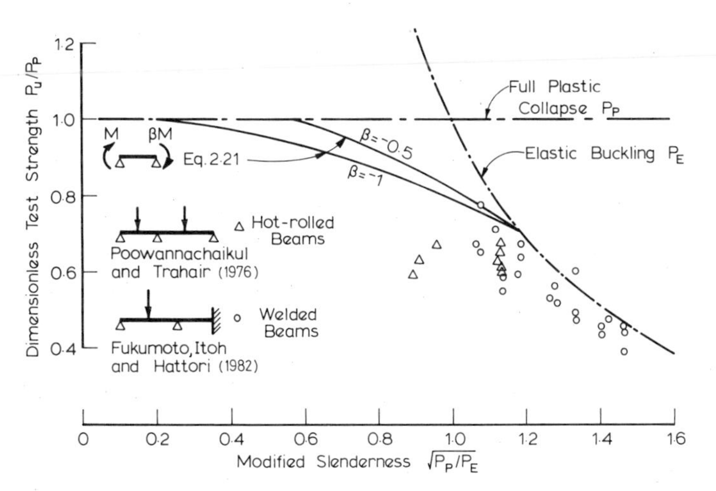

FIG. 3.6. Test results for continuous beams.

of Chapter 2. This figure suggests that the equivalent β values lie in the range $-0{\cdot}7<\beta<-0{\cdot}6$. Thus if the effects of lateral continuity between adjacent spans during inelastic buckling are ignored, then Fig. 3.6 suggests that useful tests should have modified slendernesses in the range $0{\cdot}6<\sqrt{(P_P/P_E)}<1{\cdot}1$.

3.3 DESIGN METHODS

3.3.1 Bases for Design

Design methods for laterally unsupported beams should provide a clear relationship between the strength of the beam and the major parameters necessary to describe the problem. Taking the moment capacity M_u as the most appropriate measure of the strength leads, after due consideration of the information presented earlier in this chapter as well as in Chapters 1 and 2, to

$$M_u = f(\text{beam geometry, material properties, load pattern, support conditions, geometrical and material 'imperfections'}) \tag{3.1}$$

Use of elastic and inelastic theory provides guidance on the role of the first four factors. Inelastic buckling theory can also assist in understanding the significance of some forms of material 'imperfection' such as residual stresses. However, a full appreciation of the role of all 'imperfections' can only be obtained by the careful evaluation of realistic test data. Thus both theory and experiment should be considered as complementary in deriving design methods: theory provides the correct forms of the relationships between the various controlling parameters, and experiment enables these relationships to be adjusted so that they match the observed behaviour.

Apart from the obvious requirement that design methods must provide accurate predictions of strength, it is highly desirable that this technical accuracy is not obtained at the expense of unnecessary complication. Some form of compromise between accuracy and rigour on the one hand and ease and quickness of use on the other is therefore necessary. Because basic lateral buckling theory (see eqn (1.5) of Chapter 1, for example) is normally regarded as too complex for direct use in design, it is necessary to introduce certain simplifications. Since there are several ways in which this can be done, it is possible to devise a number of approaches to the same problem. Some of these are discussed in the following sections.

3.3.2 Present Design Methods

Most present formulations for the design of beams are used either in elastic design or plastic design. In plastic design, the calculated plastic collapse loads are compared with factored loads obtained by multiplying the working loads by appropriate load factors. The slenderness of a beam designed plastically is limited to ensure that the full plastic mechanism can develop, as explained in Sections 3.2.3 and 3.2.4.

In elastic design, the working load stresses are compared with permissible stresses obtained from ultimate strength approximations by using appropriate factors of safety. For low slenderness beams, the ultimate strengths are usually taken as M_P, unless reductions are required to allow for local buckling in thin flanges. For very slender beams, the ultimate strength is usually related to the elastic buckling strengths of beams under certain loading, support and restraint conditions. The ultimate strengths of beams of intermediate slenderness are commonly approximated by empirical transitions between the plastic collapse and elastic buckling strengths. These transitions rarely account for the different residual stresses in welded beams, or for the effects of the bending moment distribution on inelastic buckling.

The present British, American and Australian methods of designing beams have been discussed by Trahair (1977). These are summarised in the following subsections.

(a) UK Method

Design rules for laterally unbraced steel beams are given in BS 449: 1969 and in BS 153: 1972. In both cases these are based (Kerensky *et al.*, 1956; Dibley, 1969) on limiting the maximum stress F_L in an initially bowed elastic beam to the yield stress F_Y. Some empirical adjustments have been made in the inelastic transition region in order to accommodate the horizontal plateau $F_L = F_Y$ necessary for stocky beams. Permissible stresses F_{bc} are obtained by dividing the limiting stresses by appropriate factors of safety. Some recognition that the maximum available strength of a stocky beam is M_P rather than M_Y is included by using a lower factor of safety against first yield of about 1·52, which when multiplied by the average shape factor for I-beams of 1·15 gives a factor of 1·75 against collapse.

Values of F_L and hence also values of F_{bc} are obtained directly from elastic critical stresses F_{0b} using a Perry–Robertson type of approach that is similar to the process used for struts. Figure 3.7(a) shows the conversion used for plate girders in BS 449: 1969 (a similar process is used for both

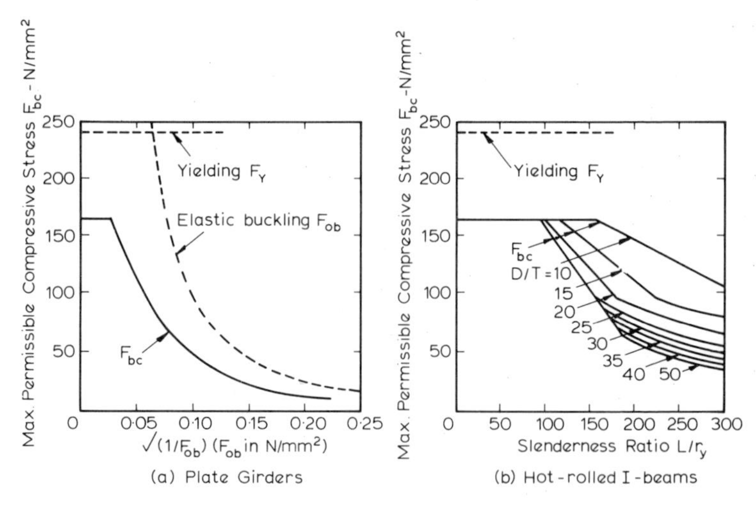

FIG. 3.7. Maximum permissible compressive stresses according to BS 449: 1969 for beams of grade 43 steel ($F_Y = 240$ N mm^{-2}).

plate girders and rolled sections in BS 153: 1972). For rolled sections only, BS 449: 1969 combines the calculation of F_{0b} and the conversion of F_{bc} into one step so that values of F_{bc} are tabulated directly in terms of the beam's effective slenderness ratio l/r_y and a cross-section parameter D/T (see Fig. 3.7(b)).

The calculation of F_{0b} is facilitated by the use of a series of approximations for the quantities appearing in eqn (1.5) of Chapter 1, as shown in Table 3.1. These enable F_{0b} to be written (after some rounding) as

$$F_{0b} = M_E/Z_x$$

$$\simeq \frac{2\,800\,000}{(l/r_y)^2}\left[1 + \frac{1}{20}\left(\frac{l}{r_y}\frac{T}{D}\right)^2\right]^{1/2} \text{N mm}^{-2} \qquad (3.2)$$

Because eqn (3.2) is conservative for many sections, the value of F_{0b} may be increased by 20% provided certain geometrical restrictions are met. The procedure for rolled sections of BS 449: 1969 automatically includes this allowance.

The influence of different forms of loading is included only in the case of cantilevers, for which a series of effective length factors is provided, or in the case of simply supported beams with top flange loading free to move sideways as the beam buckles, for which a effective length of $1{\cdot}2L$ is specfied.

TABLE 3.1
APPROXIMATIONS TO CROSS-SECTIONAL PROPERTIES

BS 449: 1969	*AISC 1969 Specification*
$Z_x = 1{\cdot}1BTD$	$Z_x = BTD[1 + (D - 2T)t/6BT]$
$I_y = B^3T/6$	$I_y = B^3T/6$
$J = 0{\cdot}9BT^3$	$J = 2BT^3/3 + Dt^3/3$
$I_w = I_yD^2/4$	$I_w = I_yD^2/4$
$B = 4{\cdot}2r_y$	$\dfrac{1 + (D - 2T)t^3/2BT^3}{[1 + (D - 2T)t/6BT]^2} = 1$
$E = 2{\cdot}5G$	$E = 2{\cdot}59G$

(*b*) *US Method*

The starting point for the AISC 1969 Specification provisions is also a simplified version of eqn (1.5) of Chapter 1. Table 3.1 lists the approximations used to arrive at an expression for the elastic critical stress F_{0b} in the form

$$F_{0b} \simeq \left[\left(\frac{130\,000}{lD/BT}\right)^2 + \left(\frac{1\,975\,000}{(l/r_T)^2}\right)^2\right]^{1/2} \text{ N mm}^{-2} \tag{3.3}$$

in which r_T is the radius of gyration about the y-axis of the compression flange plus one-sixth of the web.

Depending on the proportions of the cross-section, one or other of the terms in eqn (3.3) will usually be dominant. Conservative approximations to the permissible stress for slender beams may therefore be obtained by dividing these approximations to F_{0b} by appropriate safety factors. For the first term 1·57 is used to give

$$F_{bc} = \frac{82\,000}{lD/BT} \text{ N mm}^{-2} \tag{3.4}$$

whilst for the second term a value of 1·69 is adopted, yielding

$$F_{bc} = \frac{1\,172\,000}{(l/r_T)^2} \text{ N mm}^{-2} \tag{3.5}$$

For beams of intermediate slenderness, a transition through the inelastic range is provided, which is similar to the Specification's treatment for inelastic column buckling, i.e.

$$F_{bc} = \frac{2F_Y}{3}\left[1 - \frac{1}{(2 \times 16/9)}\frac{F_Y(l/r_T)^2}{1\,975\,000}\right] \tag{3.6}$$

This equation should be used in place of eqn (3.4) when $\sqrt{[F_Y(l/r_T)^2/1\,975\,000]} < 4/3$. In addition, since the maximum permissible stress F_{bc} may be taken as the higher of the values obtained from eqns (3.4) and (3.5), both eqns (3.4) and (3.6) should also be checked in the case when $\sqrt{[F_Y(l/r_T)^2/1\,975\,000]} < 4/3$ and the higher figure used, subject to an upper limit of

$$F_{bc} = 0{\cdot}60 F_Y \tag{3.7}$$

Equation (3.7) corresponds to a nominal factor of safety on M_Y of $1/0{\cdot}6 \simeq 1{\cdot}67$.

For short beams ($l/B < 200/\sqrt{F_Y}$ and $l/B < 138\,000T/DF_Y$) the 0·60 in eqn (3.7) is replaced by 0·66, corresponding to a factor of safety against M_P of $1/(0{\cdot}66/1{\cdot}1) \simeq 1{\cdot}67$ if a low value of 1·1 is assumed for the shape factor.

The application of this method to a beam for which $F_Y = 248$ N mm^{-2} and $r_T D/BT = 9{\cdot}23$ is illustrated in Fig. 3.8. It can be seen that it requires repeated calculation because, except for very long or very short beams, it is not possible to anticipate which formula will govern. Graphs provided in the Specification alleviate this problem for standard beams.

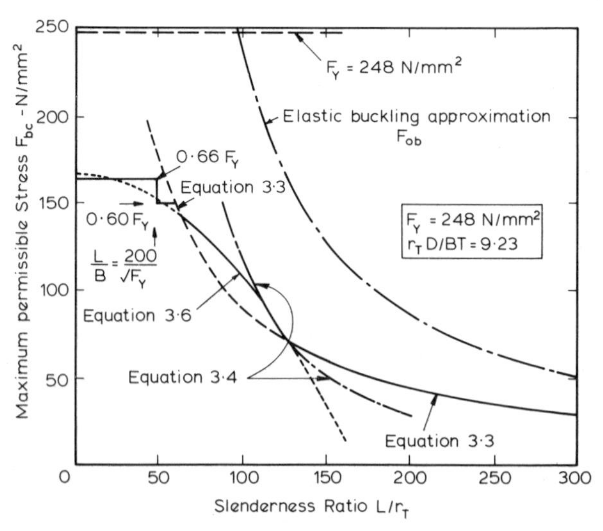

FIG. 3.8. Maximum permissible stresses of the AISC 1969 Specification.

Allowance for moment gradient loading is through the use of equivalent uniform moment factors of the type given as eqn (1.11) (see Chapter 1) but with an upper limit of 2·3.

(*c*) *Australian Method*

AS1250-1981 uses a set of three semi-empirical expressions, corresponding to the three regions of beam behaviour, to link the permissible stress F_{bc} directly to yield stress F_Y and elastic critical stress F_{Ob}. These are:

$$F_{bc}/F_{Ob} = 0{\cdot}55 - 0{\cdot}10 F_{Ob}/F_Y \qquad \text{(for slender beams)} \qquad (3.8)$$

$$F_{bc}/F_Y = 0{\cdot}95 - 0{\cdot}50(F_Y/F_{Ob})^{1/2} \qquad \text{(for beams of intermediate slenderness)} \qquad (3.9)$$

For stocky beams, a linear transition is used between values of F_{bc} of $0{\cdot}60F_Y$ and $0{\cdot}66F_Y$, depending on compression flange width to thickness ratio. This implies a safety factor of $1/0{\cdot}60 \simeq 1{\cdot}67$ compared with a value of $1/0{\cdot}55 \simeq 1{\cdot}82$ in the case of long beams. The comparison with test data shown in Fig. 3.9 indicates that for beams of intermediate slenderness, the load factors associated with eqn (3.9) may be somewhat lower.

Use of eqns (3.8) and (3.9) requires a knowledge of the elastic critical stress F_{Ob}. Two methods are permitted: (i) use of an equation similar to eqn (3.2), or (ii) use of values obtained directly from an elastic buckling

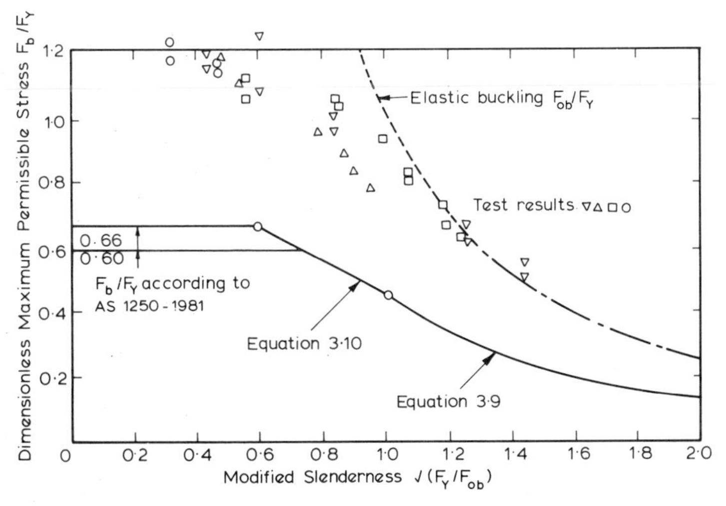

FIG. 3.9. Maximum permissible stresses of the AS 1250–1981.

analysis. In the case of the second method, recourse may be made to published data or computer-based methods of the type discussed in Chapter 1, thereby enabling an accurate allowance to be made for the precise form of loading and type of support conditions within the framework of the code approach.

3.3.3 Recent Developments in Design Methods

(a) Statistical Assessment of Test Results

Statistical assessments have been made of early buckling tests by Fukumoto and Kubo (1977*a*, *b*, *c*). Their values of the mean strengths (m) and mean minus two standard deviations ($m - 2s$) are shown in Figs. 3.3, 3.4 and 3.5. It can be seen from Figs. 3.3 and 3.4 that the ($m - 2s$) values are quite close to the lowest values reported, and it can be concluded that these values form approximate lower bounds suitable for design.

Formulations for these lower bounds have been developed. These often take the form of

$$M_{u0}/M_P = 1/(1 + (M_P/M_E)^n)^{1/n} \tag{3.10}$$

in which n is chosen to match the curve to the test results. Fukumoto and Kubo (1977*a*, *c*) suggested that $n = 1{\cdot}5$ defines a lower bound for the hot-rolled beam tests which they reviewed, and $n = 1$ for the welded beam tests. Curves for these values of n are shown in Figs. 3.6 and 3.10 and good

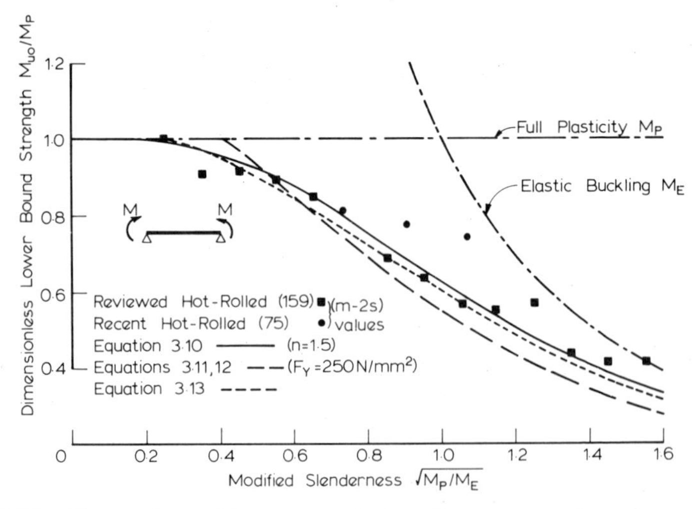

FIG. 3.10. Comparison of lower bound strength approximations for uniform bending.

agreement with the $(m - 2s)$ strength values is demonstrated. Values of n of 2·5 and 2·0 respectively for hot-rolled and welded sections have been adopted by the European Convention for Construction Steelwork (ECCS, 1978) in their Recommendations. This has led to the introduction of similar proposals by a number of European countries such as Switzerland and West Germany (SSRC/ECCS/CRCJ/CMEA, 1981).

Also shown in Figs. 3.5 and 3.6 are the $(m - 2s)$ strength values for recent tests (Fukumoto *et al.*, 1980; Fukumoto and Itoh, 1981). The higher values for these tests on beams with low geometrical imperfections and flange residual stresses have already been discussed.

While the formulation of eqn (3.10) gives satisfactory agreement with test results in the inelastic and elastic buckling ranges, it has a serious defect. This is that it precludes the use of the full plastic capacity, since the value calculated for the strength M_n is always less than M_P, except at zero slenderness (when $M_E = \infty$). This defect has been eliminated in a proposal for a revised British Code reported by Taylor *et al.* (1974), in which the design ultimate strength M_{u0} of a hot-rolled beam is calculated from

$$(M_E - M_{u0})(M_P - M_{u0}) = \eta M_E M_{u0} \tag{3.11}$$

in which

$$\eta = 0{\cdot}007\pi\sqrt{(E/F_Y)}(\sqrt{(M_P/M_E)} - 0{\cdot}4) \nless 0 \tag{3.12}$$

which defines a slenderness range of $0 < \sqrt{(M_P/M_E)} < 0{\cdot}4$ in which $M_{u0} = M_P$, as shown in Fig. 3.10. However, the formulation does not fit the $(m - 2s)$ strength values of the hot-rolled tests reviewed at all well, possibly because the chosen plastic range limit of $\sqrt{(M_P/M_E)} = 0{\cdot}4$ is too high for beams in near-uniform bending (it appears from Fig. 2.16 of Chapter 2 that 0·2 is a better value of this limit for such beams, while eqn (2.22) of Chapter 2 suggests a value of 0·17).

An improved formulation for beams in uniform bending is given by

$$M_{u0}/M_P = 0{\cdot}6\{\sqrt{[(M_P/M_n)^2 + 3]} - M_P/M_0\} \ngtr 1{\cdot}0 \tag{3.13}$$

and this is shown in Fig. 3.10. It can be seen that this gives reasonable agreement with the $(m - 2s)$ strength values of hot-rolled beams, with a plastic limit of $\sqrt{(M_P M_0)} = 0{\cdot}26$.

(b) Welded Beams

The $(m - 2s)$ strength values of welded beams shown in Fig. 3.5 for both the early and recent tests are significantly lower than those of the corresponding hot-rolled beams. It has been proposed (Fukumoto and

Kubo, 1977*a*, *c*) that this should be allowed for by using the lower value of $n = 1$ in eqn (3.10), instead of $n = 1{\cdot}5$ for hot-rolled beams. The effect of this is shown in Fig. 3.5.

A different method has been proposed by Taylor *et al.* (1974), in which the yield stresses of welded beams are reduced by $20\,\mathrm{N\,mm^{-2}}$, and the design strengths M_{u0} calculated from eqns (3.11) and (3.12). Because this produced only small decreases in the strengths calculated for beams of intermediate slenderness, it was subsequently replaced by the use of a different design curve, obtained by redefining η for welded sections as

$$\begin{aligned}\eta &= 0{\cdot}0056\pi\sqrt{(E/F_Y)}\\ &\not> 0{\cdot}016\pi\sqrt{(E/F_Y)}(\sqrt{(M_P/M_E)} - 0{\cdot}4)\\ &\not< 0{\cdot}007\pi\sqrt{(E/F_Y)}(\sqrt{(M_P/M_E)} - 0{\cdot}4)\\ &\not< 0{\cdot}0\end{aligned} \tag{3.14}$$

This provides a better description of the test data.

(*c*) *Bending Moment Distribution*

Several proposals have been made for incorporating into design codes allowances for the effect of the moment distribution on inelastic buckling. In one of these, Nethercot and Trahair (1976) suggested that the design strength M_u should be calculated directly from the buckling strength M_c, in which M_c is the appropriate elastic (M_E) or inelastic (M_I) buckling moment for the actual bending moment distribution. Their relationships for this calculation are

$$M_u = M_P \tag{3.15}$$

for stocky beams ($\sqrt{(M_P/M_c)} < 1{\cdot}0$),

$$M_u = M_c(1{\cdot}57M_c/M_P - 0{\cdot}57) \tag{3.16}$$

for beams of intermediate slenderness ($1{\cdot}0 < \sqrt{(M_P/M_c)} < 1{\cdot}1$), and

$$M_u = M_c(0{\cdot}95 - 0{\cdot}27M_c/M_P) \tag{3.17}$$

for slender beams ($1{\cdot}1 < \sqrt{(M_P/M_c)}$).

These equations, which were chosen to provide an approximate lower bound to selected test results, have been shown by Fukumoto and Kubo (1977*b*) to be close to the lower bounds of the hot-rolled beam tests that they reviewed. Later, Nethercot and Trahair (1977) made slight modifications to these equations.

Other design methods have been proposed which allow for the effects of moment distribution on inelastic buckling. Taylor *et al.* (1974) suggested that for beam segments loaded only at points of lateral restraint, the moment modification factor m normally used in eqn (1.10) of Chapter 1 to approximate the elastic buckling moment M_E should be used instead to calculate the design strength from

$$M_u = mM_{u0} \not> M_P \tag{3.18}$$

in which M_{u0} is the design ultimate strength for uniform bending obtained from eqns (3.11), (3.12) and (3.14). Yura *et al.* (1978) proposed to use eqn (3.18) with their straight line (with span length) approximation for the inelastic buckling of beams in uniform bending.

This method (eqn (3.18)) can also be used in conjunction with the improved formulation for beams in uniform bending given by eqn (3.13). Curves of M_u/M_P calculated in this way are compared with those of M_I/M_P approximations (eqn (2.21) of Chapter 2) for inelastic buckling in Fig. 3.11. It can be seen that these curves are similar to that for uniform bending (eqn (3.13)), which agreed closely with the $(m - 2s)$ strength values of the reviewed tests, and that they define plastic limits (at which $M_u = M_P$) which are in reasonable agreement with those predicted by inelastic buckling analyses (Fig. 2.16 of Chapter 2).

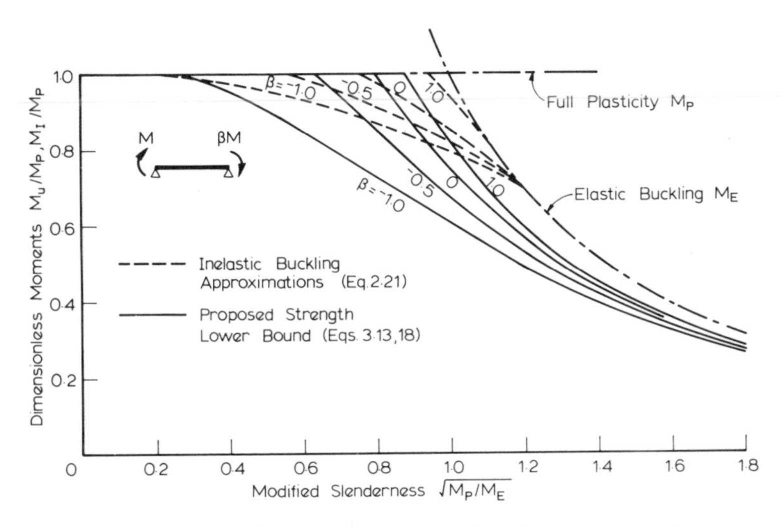

FIG. 3.11. Lower bound strength approximations for unequal end moments.

(*d*) *Limit States Design*

During the last decade, the development commenced of a new and more rational type of design method, called 'Limit States Design' (LSD) or 'Load and Resistance Factor Design' (LRFD). In this method, the various design criteria are expressed in a common form such as (Galambos and Ravindra, 1974):

$$\phi R_n > \sum \gamma_k Q_k \tag{3.19}$$

in which R_n is the nominal resistance, Q_k are the nominal load effects, and ϕ and γ_k are appropriate resistance and load factors. In the LRFD approach these factors are determined, after gathering statistical data on the actual resistances R and load effects Q, by using an approximately constant safety index β defined by

$$\beta = \frac{(\ln (R/Q))_{\text{mean}}}{(\ln (R/Q))_{\text{standard deviation}}} \tag{3.20}$$

This method is intended to give greater consistency in the reliability of the structures designed than is possible with the present working stress or ultimate load design methods, to allow the use of target reliabilities which are appropriate to the construction, use and importance of the structure, and to provide a common basis for designing a wide range of structures of different materials and uses.

Ravindra and Galambos (1978) examined the US working stress design methods for steel members, and found that they correspond to a safety index of $\beta = 3$ approximately. Yura *et al.* (1978) recommended that the nominal bending resistance R_n for beams should be taken as either the full plastic moment M_P or the inelastic and elastic buckling moments M_I and M_E, as appropriate. They then found that the ϕ factors required to achieve a constant safety index of $\beta = 3{\cdot}0$ varied from 0·78 (for beams of intermediate slenderness) to 0·84 (for slender beams) and 0·89 (for beams of low slenderness). For simplicity they recommended the adoption of the constant value of $\phi = 0{\cdot}86$.

The variability of the calculated ϕ factors to some extent reflects the fact that the buckling moments M_I and M_E are greater than the strengths M_n of real beams which have geometric imperfections. It is therefore suggested that the use of the proposed strength lower bound of eqns (3.13) and (3.18) (Fig. 3.10) will give more nearly uniform ϕ factors than the buckling moments M_I and M_E.

(*i*) *UK methods*. Equation (3.11) has been adopted as the basis for both the new UK steel bridge code BS 5400: Part 3: 1982 and the draft building code BS 5950. Whilst the latter retains eqns (3.12) and (3.14) for η, only one curve for all types of section defined by (Nethercot, 1981)

$$\eta = 0{\cdot}005\sqrt{(\pi^2 E/355)}(\sqrt{(M_P/M_E)} - 0{\cdot}6) \tag{3.21}$$

is given in BS 5400: Part 3: 1982.

Evaluation of eqn (3.21) is simplified through the introduction of a lateral slenderness ratio λ_{LT} (Taylor *et al.*, 1974; Nethercot and Taylor, 1977; Nethercot, 1981), defined by

$$\lambda_{LT} = \pi\sqrt{(E/F_Y)}\sqrt{(M_P/M_E)} \tag{3.22}$$

This may be evaluated from

$$\lambda_{LT} = uv\lambda \tag{3.23}$$

in which $u = [(4Z_P^2\gamma)/(A^2h^2)]^{1/4}$ is the buckling parameter, $x = 0{\cdot}566h(A/J)^{1/2}$ is the torsional index, $\gamma = (1 - I_y/I_x)$, $v = [1 + \frac{1}{20}(\lambda/x)^2]^{-1/4}$, and $\lambda = l/r_y$. Safe approximations for u:

$$\begin{aligned} u &= 0{\cdot}9 \text{ for rolled UB, UC and channel sections} \\ u &= 1{\cdot}0 \text{ for all other sections} \end{aligned} \tag{3.24}$$

which may be used in conjunction with the approximation $x = D/T$, facilitate the method's use. As an alternative, values of u and x for handbook sections are tabulated. Procedures for determining λ_{LT} for other forms of cross-section, e.g. monosymmetric I's, flats, tapered sections, etc., are provided.

Both codes are written in limit state format. However, rather than using the safety index method (Section 3.3.3(d)), they employ the partial safety factor format (CIRIA, 1968). The basis is still eqn (3.19), but the method used to assign numerical values to ϕ_n and γ_k is different (CIRIA, 1977; Flint *et al.*, 1981).

(*ii*) *Proposed US method.* Galambos (private communication) has given details of a formulation for a proposed US LRFD Code. In this, the design strength is calculated from

$$M_u = m\left[M_P - (M_P - M_r)\frac{(\lambda - \lambda_P)}{(\lambda_r - \lambda_P)}\right] \not> M_P \tag{3.25}$$

where $\lambda \leq \lambda_r$, and

$$M_u = mM_0 \tag{3.26}$$

where $\lambda \geq \lambda_r$, in which M_0 is the elastic buckling moment of a beam in uniform bending (eqn (1.5) in Chapter 1), m is the moment modification factor for beams with unequal end moments (eqn (1.11) in Chapter 1, but with a maximum value of 2.3) and

$$M_r \simeq M_Y \left(1 - \frac{0{\cdot}28}{F_Y/248}\right) \tag{3.27}$$

is the moment which would cause the nominal first yield in a beam having a residual stress of $0{\cdot}28 \times 248 = 69\ \text{N mm}^{-2}$.

The geometrical slendernesses used in eqn (3.25) are defined by

$$\left.\begin{aligned} \lambda &= l/r_y \\ \lambda_P &= 50\sqrt{F_Y/248} \\ \lambda_r &= (\lambda)_{M_0 = M_r} \end{aligned}\right\} \tag{3.28}$$

In this, λ_P sets a limit to the geometrical slenderness of beams in uniform bending ($m = 1$) which are permitted to reach M_P.

Equation (3.25) provides a linear decrease in strength with increase in (l/r_y) in the inelastic buckling range, and takes account of the increased resistance of beams under high moment gradient. For slender beams $(\lambda \geq \lambda_r)$, eqn (3.26) uses the elastic buckling resistance for the strength.

3.4 CONCLUDING REMARKS

A review has been made of the available experimental data relating to the lateral buckling of steel beams. Using this in conjunction with the theories of Chapters 1 and 2, the background to several design approaches has been presented. These cover both established permissible stress methods and the newer limit states procedures.

REFERENCES

BJHOVDE, R. (1978) The safety of steel columns. *Journal of the Structural Division, ASCE*, **104**(ST3), Proc. Paper 13607, 463–77.

CIRIA (1968) Study Committee on Structural Safety, Guidance for the drafting of codes of practice for structural safety. Technical Note No. 2. Construction Industry Research and Information Association (Aug.).

CIRIA (1977) The rationalisation of safety and serviceability factors in structural codes. Construction Industry Research Information Association, Report No. 63 (July).

DIBLEY, J. E. (1969) Lateral-torsional buckling of I-sections in grade 55 steel. *Proceedings, Institution of Civil Engineers*, **43**, 599–627.

ECCS (1978) *European Recommendations for Steel Construction*. European Convention for Constructional Steelwork (Mar.).

FLINT, A. R., SMITH, B. W., BAKER, M. J. and MANNERS, W. (1981) The derivation of safety factors for design of highway bridges. In *The Design of Steel Bridges*, ed. K.C. Rockey and H.R. Evans, Granada Publishing, St Albans, pp. 11–36.

FUKUMOTO, Y. (1981) Numerical data bank for the ultimate strength of steel structures. *Proceedings, US–Japan Seminar on Inelastic Instability of Steel Structures and Structural Elements*, Tokyo.

FUKUMOTO, Y. and ITOH, Y. (1981) Statistical study of experiments on welded beams. *Journal of the Structural Division, ASCE*, **107**(ST1), Proc. Paper 15965, 89–103.

FUKUMOTO, Y. and KUBO, M. (1977*a*) A survey of tests on lateral buckling strength of beams. *Preliminary Report, 2nd International Colloquium on Stability of Steel Structures*, ECCS–IABSE, Liège, pp. 233–40.

FUKUMOTO, Y. and KUBO, M. (1977*b*) A supplement to a survey of tests on lateral buckling strength of beams. *Final Report, 2nd International Colloquium on Stability of Steel Structures*, ECCS–IABSE, Liège, pp. 115–17.

FUKUMOTO, Y. and KUBO, M. (1977*c*) An experimental review of lateral buckling of beams and girders. *Proceedings, International Colloquium on Stability of Structures under Static and Dynamic Loads*, SSRC–ASCE, Washington, pp. 541–62.

FUKUMOTO, Y., ITOH, Y. and KUBO, M. (1980) Strength variations of laterally unsupported beams. *Journal of the Structural Division, ASCE*, **106**(ST1), Proc. Paper 15142, 165–81.

FUKUMOTO, Y., ITOH, Y. and HATTORI, R. (1982) Lateral buckling tests on welded continuous beams. *Journal of the Structural Division, ASCE*, **108**(ST10), 2245–62.

GALAMBOS, T. V. and RAVINDRA, M. K. (1974) Load and resistance factor design criteria for steel beams. Washington University, Civil and Environmental Dept, Research Report No. 27 (Feb.).

KERENSKY, O. A., FLINT, A. R. and BROWN, W. C. (1956) The basis for design of beams and plate girders in the revised British Standard 152. *Proceedings, Institution of Civil Engineers*, **5**(2), 396–461.

NETHERCOT, D. A. (1981) Design of beams and plate girders: treatment of overall and local flange buckling. In *The Design of Steel Bridges*, ed. K. C. Rockey and H.R. Evans, Granada Publishing, St Albans, pp. 243–62.

NETHERCOT, D. A. and TAYLOR, J. C. (1977) Use of a modified slenderness in the design of laterally unsupported beams. *Preliminary Report, 2nd International Colloquium on Stability of Steel Structures*, ECCS–IABSE, Liège, pp. 197–202.

NETHERCOT, D. A. and TRAHAIR, N. S. (1976) Inelastic lateral buckling of determinate beams. *Journal of the Structural Division, ASCE*, **102**(ST4), Proc. Paper 12020, 701–17.

NETHERCOT, D. A. and TRAHAIR, N. S. (1977) Design rules for the lateral buckling of steel beams. *Civil Engineering Transactions, Institution of Engineers, Australia*, **CE19**(2), 162–5.

OJALVO, M. and WEAVER, R. (1978) Unbraced length requirements for steel I-beams. *Journal of the Structural Division, ASCE*, **104**(ST3), Proc. Paper 13634, 479–90.

POOWANNACHAIKUL, T. and TRAHAIR, N. S. (1976) Inelastic buckling of continuous steel I-beams. *Civil Engineering Transactions, Institution of Engineers, Australia*, **CE18**(2), 134–9.

RAVINDRA, M. K. and GALAMBOS, T. V. (1978) Load and resistance factor design for steel. *Journal of the Structural Division, ASCE*, **104**(ST9), 1337–53.

SSRC/ECCS/CRCJ/CMEA (1981) Stability of metal structures: a world view. *AISC Engineering Journal*, **18**(4), 154–96.

TAYLOR, J. C., DWIGHT, J. B. and NETHERCOT, D. A. (1974) Buckling of beams and struts: proposals for a new British code. *Proceedings, Conference on Metal Structures and the Practising Engineer, Institution of Engineers, Australia*, Melbourne (Nov.), pp. 27–31.

TRAHAIR, N. S. (1977) *Behaviour and Design of Steel Structures*, Chapman and Hall, London.

WAKABAYASHI, M. and NAKAMURA, T. (1983) Buckling of laterally braced beams. *Engineering Structures*, **5**(2), 108–18.

YURA, J., GALAMBOS, T. V. and RAVINDRA, M. K. (1978) The bending resistance of steel beams. *Journal of the Structural Division, ASCE*, **104**(ST9), Proc. Paper 14015, 1355–70.

Chapter 4

DESIGN OF I-BEAMS WITH WEB PERFORATIONS

R. G. REDWOOD

Department of Civil Engineering and Applied Mechanics, McGill University, Montreal, Canada

SUMMARY

In this chapter methods of analysis, and their adaptation for design purposes, are described for I-beams containing large web openings. Attention is directed to relatively stocky webbed beams typical of beams in building structures. Elastic and ultimate strength analyses are treated separately, with emphasis on the latter. Circular and rectangular openings which may be centred at the mid-depth of the beam, or eccentric to it, are considered, and analyses without reinforcement and with horizontal bar reinforcement are described. This reinforcement is structurally efficient and economically attractive. The strength of composite beams with large web openings is also treated.

NOTATION

a	Half length of opening
$\bar{a}$	Depth of concrete compression zone
b	Total flange width of steel section
b_e	Effective width of composite beam web
b_r	Width of reinforcing bar
d	Overall depth of steel section
e	Eccentricity of mid-depth of hole above mid-depth of steel beam section
f_c'	Compression (cylinder) strength of concrete

g	Centre-to-centre distance between adjacent openings
k_1, k_2	Factors defining points of stress reversal
n_h	Number of shear connectors within length of opening
q	Ultimate resistance of one shear connector
s, s_1, s_2	Height of tee-section web, above opening, below opening
t	Flange thickness of steel section
t_r	Thickness of reinforcing plate
w	Web thickness
$\bar{y}, \bar{y}_1, \bar{y}_2$	Distances from opening edge to plastic neutral axis of tee-section
A_f	Area of one flange ($= bt$)
A_r	Area of reinforcing bars above or below the opening
A_{r0}	$aw/\sqrt{3}$
A_w	Gross area of web
E	Modulus of elasticity
F	Normal force
H	Half-height of opening
I, I_t, I_b	Moments of inertia of beam, tee-sections above and below opening
M	Bending moment
M_p, M_{pT}, M_{post}	Plastic moments of unperforated beam section, of tee-section and of web-post
M_0, M_1	Bending moments at specific points on interaction diagrams
R	Radius of circular opening
S_x	Elastic modulus of unperforated beam section
T_s	Overall slab thickness of composite beam
V	Shearing force
V_p, V_{pt}	Plastic shear forces of unperforated beam, and of top tee-section
V_t, V_b	Shearing forces carried by top and bottom sections
V_1, V_{t1}, V_{b1}	Shearing forces corresponding to point '1' on interaction diagrams
α, β, γ	Geometric parameters defined in text
Γ	Ratio of maximum shear stress in flanged beam to average shear stress in web
μ	Shear connector strength ratio defined in eqn (4.40)

λ	Geometric parameter related to multiple openings
ϕ	Angle defining position around circular openings
τ	Shearing stress
σ	Normal stress
σ_y	Yield stress in tension
σ_ϕ	Tangential stress on hole edge at position defined by ϕ

4.1 INTRODUCTION

Traditional structural steel floor framing in building structures consists of beams and girders with solid webs, and these frequently create problems in that they hinder the distribution of services. To pass large ducts and pipes beneath the girders may lead to unacceptably large construction depths between storeys, and a frequently used solution is to provide openings in the webs. Such openings are sometimes required to be large: for example, with height of 60 % or more of the beam depth and length of two or three times the opening height.

The problems associated with distribution ducts and pipes in solid web beams, together with modern requirements for longer column-free spans, have led to more frequent use of trussed girders in buildings. These same factors have also led to innovative framing such as stub girders and the use of cantilever construction in floor systems. However, these systems still represent a small proportion of structures requiring services and thus there is continuing need for methods of analysis and design of web opening details.

In the absence of proven methods of design for such openings, a conservative treatment of their design was necessary. With the development of sophisticated building service systems and the building of many higher structures, this was leading, in the 1960s, to high costs of fabrication of reinforcements for web openings. Thus, cases in which opening reinforcement represented as much as 3 % of the weight of the structure have been reported, and naturally the related costs represent a much higher proportion of the structure cost. The use of similar services on each floor and the possibility of changing the layout or function of a building have sometimes resulted in standardisation of openings and reinforcement in many of the beams. It is clear, therefore, that optimisation of the details associated with an opening is desirable, and significant savings in cost can result if it can be established that an opening does not require reinforcement.

Openings of circular and rectangular shape are widely used. Other shapes are less frequent, although occasionally openings which are flat on the top and bottom edges and semicircular at their ends are found, and usually the corners of rectangular openings are radiused.

Except in the case of castellated beams, reinforcement around openings is used frequently. Several different types of reinforcement are shown in Fig. 4.1. The structural purpose of these varies from types (a) to (d), which predominantly enhance the primary and local flexural strength, to type (j), which predominantly enhances the shear-carrying capacity of the web. Both flexural and web shear capacities may be improved by combination of these types or by arrangements such as types (g) to (i).

Selection of a particular reinforcement type is influenced by a number of

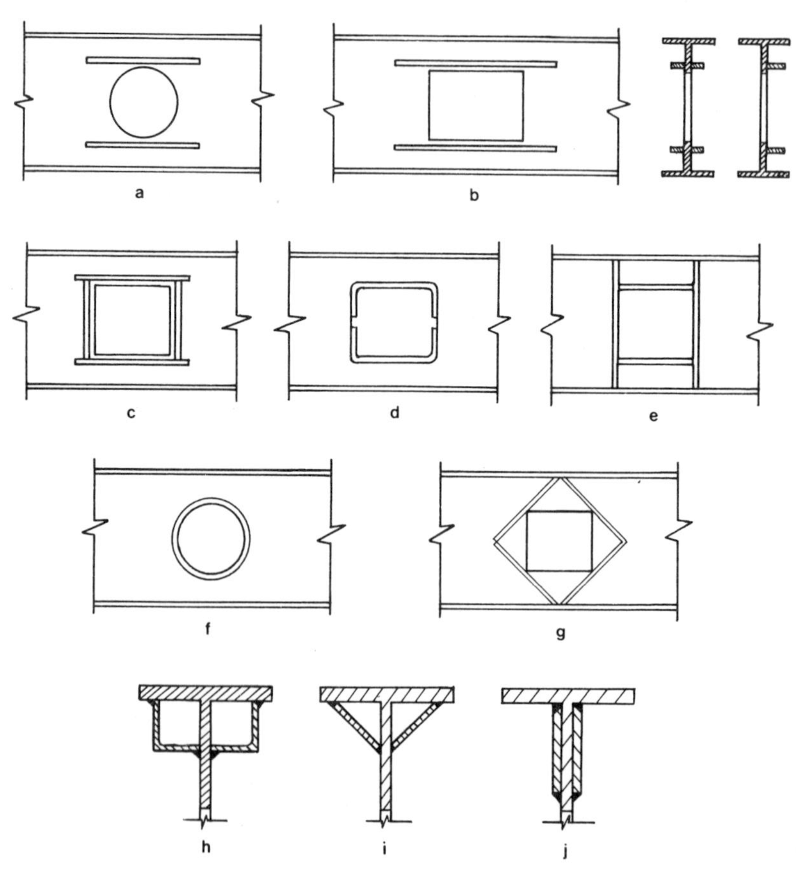

FIG. 4.1. Reinforcement of openings.

factors. The type of loading, specifically the relative importance of bending moment and shearing force, will determine whether the reinforcement should be predominantly of the flexural or shear type. Other factors influencing the choice of reinforcement type are the ease of fabrication and the degree of confidence with which the analysis can be carried out.

In this chapter, methods of analysis for beams containing large web openings are described, the emphasis being on the analysis of webs with one or two isolated openings. A considerable amount of research has been directed towards the behaviour of steel beams with web openings, and studies on steel–concrete composite beams are currently in progress. It is intended herein to provide a critical survey of existing information provided by these studies, and to emphasise those results which have particular relevance to design. Since two earlier surveys (McCormick, 1972; Redwood, 1973) were published, considerable information has become available, particularly relating to eccentric and reinforced openings and to composite beams.

Elastic and ultimate strength analyses are treated separately. In a web which remains elastic, an opening produces a stress-raising effect under all types of loading and critical stresses will usually occur at the edge of the opening. Resulting stress concentration factors are found to increase as shear-to-moment ratios increase and are sensitive to the size of the opening relative to the beam depth, and to the shape of the opening.

The corners of rectangular openings are usually radiused to reduce stress magnitudes, but it is seldom feasible to design these details to avoid local plastic deformations. Opening edge stresses are also affected by the uniformity of the cut edge of the opening, which largely depends upon the method of cutting. If the ductility of the material is relied upon to accommodate these stress raisers, then an elastic stress analysis may be relevant in design. Elastic analysis will also be appropriate if fatigue loading exists, and in the case of circular openings, edge stresses will in many cases be low enough for a factor of safety against yielding to be used as a design criterion. In view of these possible applications of elastic analysis, relevant methods are discussed in Section 4.2.

Analysis and design based upon ultimate strength has been emphasised in much of the work relating to steel beams with web openings. This is logically a satisfactory approach in view of the ductility of the material, the stress-raising effects of notches and corners and the residual stresses induced by the attachment of reinforcement. Stress redistribution as a result of inelastic action frequently results in very large ultimate strengths compared with initial yield loads. Ultimate strength methods of analysis are therefore described in detail in Section 4.3.

4.2 ELASTIC BEHAVIOUR

4.2.1 Stress Analysis

The literature on elastic stress analysis of beams with web openings is extensive. Methods of analysis may be conveniently described in three categories: (a) analytical or numerical solutions of the differential equations of the theory of elasticity; (b) numerical solutions by discrete element techniques; and (c) approximate methods based on assumptions which permit various parts of the beam to be dealt with by elementary beam theory. Under these classifications, significant contributions in the literature will now be described.

(a) Theory of Elasticity

Many of the early analyses of stresses around openings were concerned with openings in large plates subjected to uniaxial or biaxial uniform tension or compression, with applications particularly relevant to aeronautical and ship structures. Many analyses of this type are described by Savin (1961), who also provides a number of solutions for openings of elliptic, rectangular and equilateral triangular shape in rectangular sectioned beams subjected to bending and shear, and includes cases in which the openings are reinforced by peripheral elastic rings. These solutions are for openings assumed to be small in size compared with the depth of the beam, since they are based on plane stress analysis of large plates in which the region around the opening is conformally mapped on to a unit circle. This approach makes use of complex stress functions as used by Muskhelishvili. Deresiewicz (1968) has pointed out some of the restrictions of the results given by Savin and has presented mapping functions permitting more general shapes of opening to be analysed. Photoelastic results show good agreement for circular openings in rectangular sectioned beams in pure bending when the opening depth is 60 % of the beam depth, and in bending and shear when the depth ratio is 50 %.

Gibson and Jenkins (1956) obtained stress functions directly for the case of a centrally located circular opening in a simply supported beam with load applied directly over the opening. Experiments were performed on rectangular and I-shape beams, and satisfactory agreement was obtained with the predicted stresses on the side of the opening farthest from the applied load. Because of the location of opening and load, there was zero shear force at the vertical opening centreline, and although shear force existed either side of this line, the situation was essentially one of pure bending. This is substantiated by comparison of the stress distributions with more recent results (Bower, 1966*a*; Heller *et al.*, 1962; Savin, 1961).

Significant contributions to the treatment of elliptic and rectangular openings with rounded corners in plates and/or flanged beams have been made in a series of papers from the David Taylor Model Basin (Brock, 1958; Heller, 1951, 1953; Heller *et al.*, 1958, 1962), by Bower (1966*a*, *b*) and by Gotoh (1975*a*). All of these papers treat the problem using the complex variable methods of Muskhelishvili, and many are limited to the treatment of small openings; nevertheless, in many cases useful practical results are obtainable for openings up to 50% of the beam depth. Heller *et al.* (1962) provide useful results in the form of explicit expressions for the tangential normal stresses at any point on the edge of an opening. The opening may be circular, elliptical or rectangular with any desired corner radius, and results are given in graphical form for very small openings with a wide range of variables (it should be noted that Fig. 3 of this reference is in error). Additionally, these results together with those for a uniformly stressed plate (Heller *et al.*, 1958) give a solution for edge stresses around an opening located eccentrically with respect to the beam neutral axis. For a circular opening, as shown in Fig. 4.2, the tangential edge stress σ at a point defined by ϕ can be written:

$$\sigma_{\phi} = \frac{M}{S_x}\left[\left(\frac{2R}{d}\right)(\sin\phi - \sin 3\phi) + 2\left(\frac{e}{d}\right)(1 - 2\cos 2\phi)\right] + \frac{V}{A_w}\left[\left(4\Gamma - \frac{2e^2A_w}{I}\right)\sin 2\phi - \frac{A_wRe}{I}(\cos\phi - 3\cos 3\phi)\right] \quad (4.1)$$

where ϕ = the angle measured from the horizontal through the opening centre, Γ = the ratio of maximum to average nominal shear stress in the unperforated beam, and other symbols are defined in the Notation. The

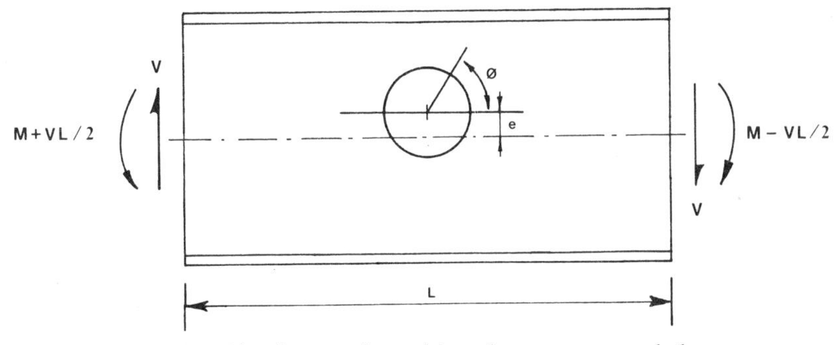

FIG. 4.2. Circular opening subjected to moment and shear.

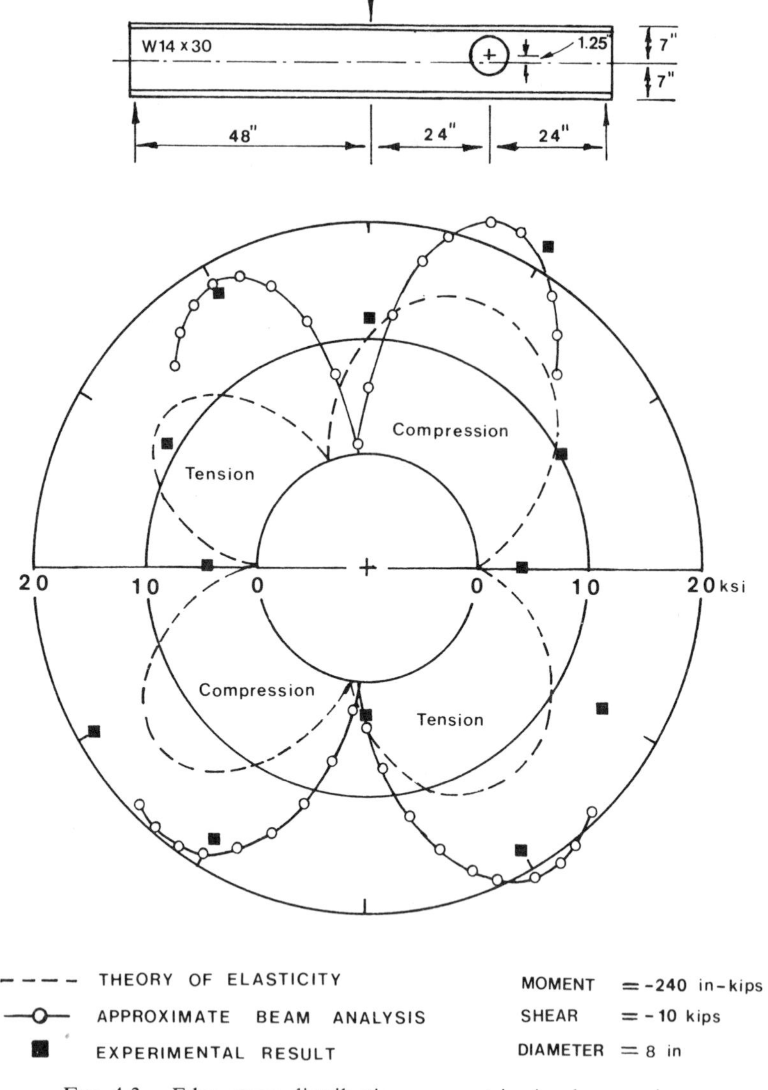

FIG. 4.3. Edge stress distributions: eccentric circular opening.

tangential stress variation around an eccentric circular opening is shown in Fig. 4.3.

Bower (1966*a*) has considered stresses throughout the web and not just at the edge of the opening, and by comparing the stress resultants with equilibrium requirements has established the size of opening above which serious errors will arise. He suggests that for the range of shear-to-moment ratios appropriate to building construction, an opening depth ratio of about 50% should be amenable to the elasticity solution, and this was confirmed by a series of tests on W410 × 54 steel beams (Bower, 1966*b*) (this is a wide-flange shape with nominal depth 410 mm and mass of 54 $kg\,m^{-1}$). Bower clarifies the relative significance of stresses due to bending moment and shearing force, and demonstrates that it is the stresses due to shearing force which suffer significant error as the depth ratio increases. This analysis forms the basis for design interaction diagrams (Bower, 1968*a*).

Gotoh (1975*a*) provides a general method for the analysis of openings in finite plates by using a combination of the complex variable method and point-matching so that the outer boundaries of the web, as well as the opening edge, can be mapped. Compared with Bower's analysis, the resulting stresses show improved comparison with test results for circular openings with diameter 50% of the beam depth. Hole edge stresses in castellated beams under pure bending have also been found by this method (Gotoh, 1975*b*).

The stresses in square and infinitely long panels in pure shear and with centrally located circular openings have been studied by Wang (1946) and Wang *et al.* (1955). Openings with diameters up to 50% of the width of the panel, both with and without reinforcement, were studied theoretically. These results were subsequently used by McCutcheon *et al.* (1963), combined with a solution for pure bending, to give the stresses around a hole under any combination of moment and shear. These results give slightly higher stresses than the values given by Heller *et al.* (1962) and Bower (1966*a*) but less than those given by Gotoh (1975*a*). Wang's results have also been used by Rockey *et al.* (1969) to verify a finite element analysis of shear panels.

Numerical solution of the differential equation governing plane stress behaviour of plates by the finite difference method was attempted by Mandel *et al.* (1971) for castellated beams. Satisfactory agreement between the analysis and experimental results was obtained in the web post and at the centreline of the opening, but theoretical results were not given for stresses near the opening corners.

(b) Analysis Using Finite Elements

Finite element analysis has been used in a number of studies to investigate conditions not amenable to solution by the methods outlined above. In most cases plane stress analysis has been used, with flanges and reinforcing plates modelled by increased element thicknesses. A summary of several studies based on this type of model is given by Redwood (1971). Elastic stresses around single, unreinforced, non-eccentric openings were verified by comparison with experimental and theory of elasticity results for openings with diameter about one-half of the hole depth. A study of various spacings of two circular openings showed that while the degree of the stress-raising effect of an adjacent opening is dependent upon the shear-to-moment ratio (the interaction increasing with this ratio), a centre-to-centre spacing of two diameters is sufficient to eliminate any interaction effects at all practical shear-to-moment ratios. These results were obtained for openings of diameter 57% of the beam depth, and were confirmed experimentally.

Studies of three types of reinforcement were also carried out. These correspond to types (a), (f) and (g) shown on Fig. 4.1, except that the type (g) did not extend up to the flange, and the openings in each case were circular. Horizontal reinforcement was found to be effective in reducing stresses only for loading with very little shearing force. (This result should not be interpreted as applying also to inelastic analysis, for which stress redistribution enables this type of reinforcement to be effective at all moment-to-shear ratios.) Inclined reinforcement is effective in reducing opening edge stresses when shear is present, and the higher the shear-to-moment ratio, the more effective the reinforcement. However, high stress concentrations occur at the reinforcement ends, and fabrication difficulties can arise with this type if the reinforcement plates intersect. Of the three types considered, circumferential reinforcement is the most effective at all shear-to-moment ratios. It was found experimentally, however, that the normal stress distribution across the reinforcement width was far from uniform owing to transverse bending of the curved reinforcement. Assumption of a reduced effective width, as given, for example, by Seely and Smith (1967) in the finite element analysis, leads to improved results.

Finite element analysis has been used to investigate the ultimate strength of unreinforced openings by Uenoya and Ohmura (1972), and reinforced eccentric openings by Cooper *et al.* (1977). Incremental loading provided information on the spread of yielding, and good agreement with ultimate load tests was obtained.

Finite element analysis has proved to be a useful research tool; its use as a design tool is obviously limited to analysis of unusual conditions and especially critical structures. The studies described above suggest that it can be used most effectively if simple plane stress elements are used in relatively fine meshes. Also, in order to simulate the continuity of the beam adequately, a zone equal in length to at least 2·5 times the opening height should be modelled.

(*c*) *Approximate Analysis*

By assuming points of contraflexure within the sections of the beam above and below the opening, the local stresses in these sections due to the shearing forces they carry can be found using engineering bending theory. These results can be added to those caused by the primary bending moment which should be based on the moment of inertia of the net section through the hole. This has been called the Vierendeel approach (Bower, 1966*b*) and has been applied to rectangular holes at mid-depth of the beam (Segner, 1964; Bower, 1966*a*), eccentric to the beam axis (Frost, 1971; Douglas and Gambrell, 1974) and openings with horizontal reinforcing bars (Cooper and Snell, 1972). Assumptions of the points of contraflexure at the mid-length of the opening has been found to give satisfactory results, and stress concentrations at hole corners are ignored.

In the case of an eccentric rectangular opening, the total shearing force V must be distributed to the parts of the beam above and below the opening. If these shearing forces are V_t and V_b, respectively, they can be found from the following (Frost, 1971):

$$V_t + V_b = V \tag{4.2}$$

$$\frac{V_t}{V_b} = \frac{\dfrac{a^2}{3EI_b} + \dfrac{1{\cdot}2}{A_{wb}G}}{\dfrac{a^2}{3EI_t} + \dfrac{1{\cdot}2}{A_{wt}G}} \tag{4.3}$$

where the subscripts t and b refer to the sections above and below the opening respectively. I is the moment of inertia, and A_w is the web area.

Circular openings are more difficult to analyse in this way because the varying depth of the tee-sections above and below the opening complicates the behaviour under both bending moment and shear force. However, methods commonly used to analyse stresses in haunched member connections (Olander, 1953) can be used in conjunction with the assumption of contraflexure at the mid-length of the opening under pure shear (Sahmel, 1969). Unreinforced mid-depth and eccentric openings are treated in this way by Chan (1971) and Chan and Redwood (1974), who show that this approximate approach gives a good estimate of the stresses around large openings as shown in Fig. 4.3. It is also shown that the theory of elasticity solution of Heller *et al.* (1962) or Bower (1966*a*) is more accurate for small holes, and that for a given opening the largest of the two stresses obtained from the approximate solution and from the theory of elasticity solution is an appropriate value to use in design.

(d) Design Based on Elastic Stresses

A number of publications have presented design aids based on one or more of the stress analyses described in the preceding sections. These include Bower (1971), Bower *et al.* (1971), Redwood and Chan (1974), Douglas and Gambrell (1974) and Höglund and Johansson (1977). These design aids either take the form of interaction diagrams relating allowable levels of shear and normal stress, or provide equations for these stresses, generally based on Vierendeel analysis. The maximum stresses may occur at a point in the beam flange and not at the hole edge, and this is taken into account in all of these design aids.

If fatigue loading is a consideration, the stresses around circular openings can be predicted sufficiently accurately to be compared with fatigue stress limits. For rectangular openings, however, because the high corner stress concentrations are usually ignored, direct comparison of the calculated stresses with fatigue limits is not meaningful. However, Frost and Leffler (1971) have shown experimentally that the opening will not be detrimental to fatigue life if the corners are radiused to the larger value of 16 mm or twice the web thickness.

4.2.2 Deflections

Deflections of a perforated beam in the elastic range can be obtained by superposition of three parts. The first is the beam deflection due to primary bending moments, and the others arise from the effect of shear force over the length of the opening. The first component can be taken as the deflection of the unperforated beam, or a small correction due to the short length with reduced stiffness at the opening can be incorporated. The second component arises from flexure of the tee-sections over the length of the opening, and can be based on the assumption that there is no relative rotation between beam sections at the opening ends. The third component is due to deflections arising from shear strains in the tee-sections, which may be of the same order of magnitude as the second component, and again no relative rotation may be assumed. These three components are shown in

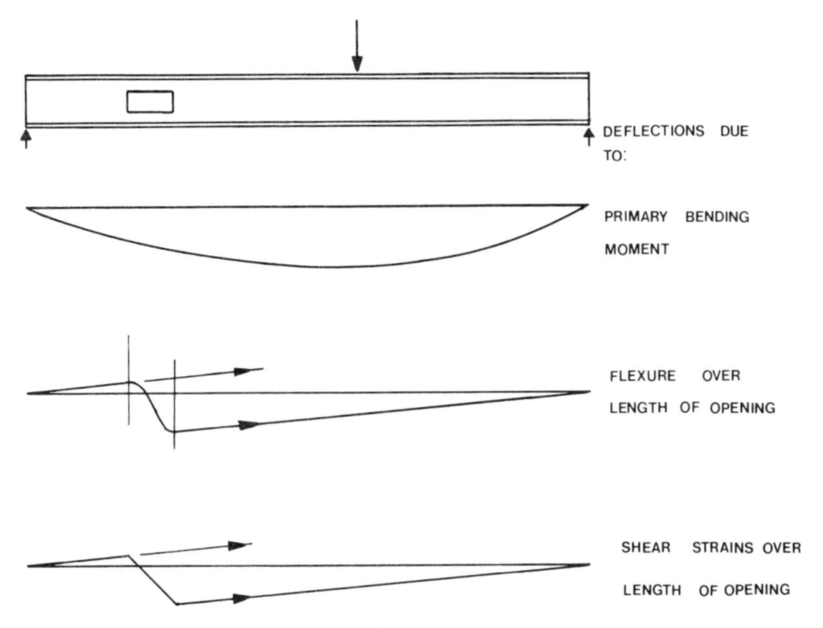

FIG. 4.4. Deflection components.

Fig. 4.4. McCormick (1972) and Dougherty (1980) provide detailed discussion of the calculation of deflection.

Circular openings with diameter up to 60 % of the beam depth have been found to cause very little additional deflection, and it has been suggested that beams with one or two isolated circular openings need no special attention. Isolated rectangular openings will seldom cause significant additional deflections in a beam. However, local deflections at an opening may vary sufficiently so that the effect on supported floor slabs and floor finishes should be considered.

4.3 ULTIMATE STRENGTH ANALYSIS

4.3.1 Introduction

The ultimate strength of perforated steel beams subjected to static loading can be predicted using plastic analysis, and this approach has been used by numerous investigators. The earliest analysis of this type appears to be that of Worley (1958), who considered rectangular, elliptical and diamond-shaped openings at the mid-depth of flanged beams. Perfectly plastic material was considered and the influence of normal and shearing forces on the values of the plastic moments was ignored. Collapse mechanisms involving four plastic hinges located as shown in Fig. 4.5(a) were assumed, and the hinge locations were varied independently in order to obtain the minimum collapse load in this upper bound solution. In the case of the rectangular opening, the minimum collapse load corresponds to hinges located at the opening corners as shown in Fig. 4.5(b), and in this case the

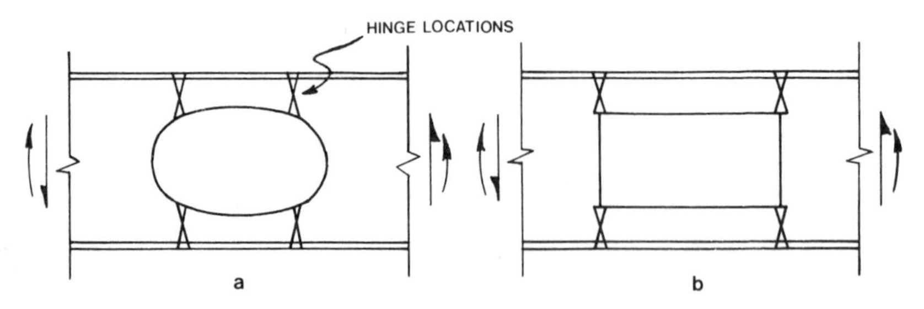

FIG. 4.5. Plastic hinge locations.

plastic moments, M_{pT}, at the four hinge locations are identical if local flexural stresses only are considered. The shearing force which will cause collapse, V, can be obtained from the virtual work equation as

$$V = \frac{2M_{pT}}{a} \tag{4.4}$$

where a is the half-length of the opening.

This method has been applied to castellated beams by Halleux (1967) and by Hope and Sheikh (1969) who, in addition, included the influence of normal force due to the primary bending moment on the plastic moments M_{pT}. Hosain and Speirs (1973) show that neglect of the normal force components can lead to significant overestimates of the strength, and conclude that neglect of shearing force effects on the plastic moment is not significant for castellated beams. In the case of isolated web openings, the shear effects may be of greater importance, and in several studies, some of which will be referred to later in greater detail, shear effects have been included in addition to normal forces and bending moments (Bower, 1967, 1968*b*; Redwood, 1968*a*; McCormick, 1972).

These latter analyses are generally lower bound solutions, that is, equilibrium is satisfied and the yield criterion is not violated throughout the beam. In some cases this is not strictly so, since while equilibrium of stress resultants is satisfied, locally, equilibrium is not. For rectangular and tee-shaped end-loaded cantilevers, Sherbourne and Van Oostrom (1972) have explored the differences between a 'stress-resultant' theory and a 'non-local' theory in which equilibrium is rigorously satisfied and as much of the section as possible attains the yield condition. Differences between the results of the two theories are significant for the tee-section cantilever, and the rigorous theory is then developed to produce moment–shear–axial force interaction diagrams for use in the analysis of the various parts of a castellated beam. Reasonable agreement with test results is demonstrated (Van Oostrom and Sherbourne, 1972).

Finite element analysis of regions of beams containing isolated openings has, on the other hand, been used to verify results of theories in which equilibrium of stress resultants only is satisfied. Excellent agreement between finite element results, plastic analysis based upon equilibrium of stress resultants, and tests is demonstrated by Cooper *et al.* (1977) for beams with eccentric reinforced rectangular openings. Other comparisons have been made by Redwood and Uenoya (1979) and show good agreement for unreinforced rectangular openings.

In view of these results, application of stress-resultant theories is examined in the following sections. The analysis of a single mid-depth web opening which considers interaction of moment, normal and shearing force in the tee-sections is first considered, followed by simplification of the results into equations suitable for design. Subsequently, more complicated cases which can be dealt with by extensions of this same analysis are described: these include eccentric openings and reinforced openings.

4.3.2 Openings without Reinforcement

(a) Mid-Depth Openings

The beam shown in Fig. 4.6 will be analysed to determine its plastic resistance in the region of the opening. Equilibrium of stress resultants near each of the hole corners is established and at the same time, based upon perfectly plastic material, full yield of these cross-sections is assumed. These locations correspond to the hinge positions shown in Fig. 4.5(b). The method follows Redwood (1968*a*, 1978*a*).

Considering the part of the beam above the opening, stress distributions at sections 1 and 2 are assumed as shown in Fig. 4.7. At section 1, where the bending moment is greatest, the normal stress reverses sign at a point within the depth of the tee-section, the position of which varies according to the value of the moment-to-shear ratio. Two cases must be considered according to whether this point is in the flange or in the web, and these

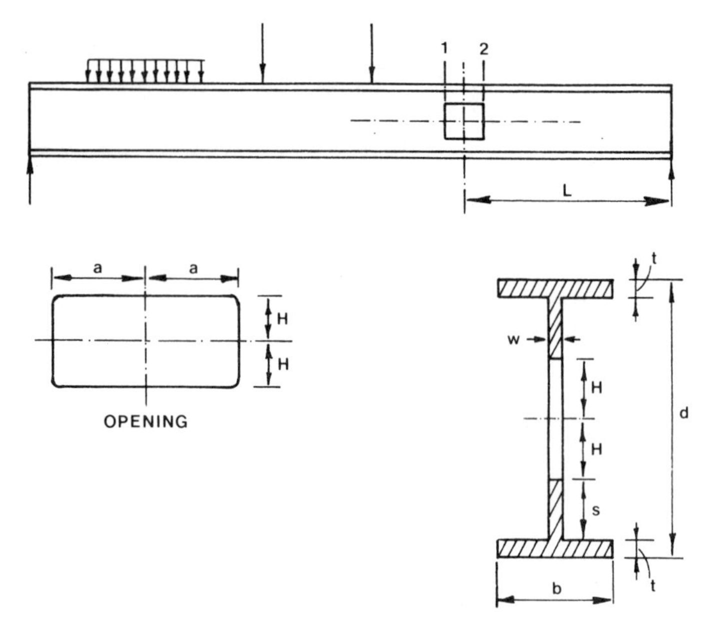

FIG. 4.6. Beam and opening configurations.

correspond to low and high values of the moment-to-shear ratio respectively. At the other end of the opening, section 2, the point of stress reversal is always in the flange.

For equilibrium:

$$M = F\bar{y}_1 + F\bar{y}_2 + 2FH \tag{4.5}$$

$$Va = F\bar{y}_1 - F\bar{y}_2 \tag{4.6}$$

where M and V are the bending moment and shearing force which are acting at the centreline of the opening at failure. The values of F, $F\bar{y}_1$, etc., can be found by integration of the assumed stresses.

For the case of low moment-to-shear ratio, Fig. 4.7(b), at section 1:

$$F = bt\sigma_y(2k_1 - 1) - sw\sigma \tag{4.7}$$

$$F\bar{y}_1 = -0{\cdot}5s^2w\sigma + bt\sigma_y[2k_1(s+t) - (s+0{\cdot}5t) - tk_1^2] \tag{4.8}$$

where $s = 0{\cdot}5d - H - t$, and σ_y is the yield stress.

At section 2, again for low moment-to-shear ratio:

$$F = sw\sigma + bt\sigma_y(1 - 2k_2) \tag{4.9}$$

$$F\bar{y}_2 = 0{\cdot}5s^2w\sigma - bt\sigma_y[2k_1(s+t) - (s+0{\cdot}5t) - tk_2^2] \tag{4.10}$$

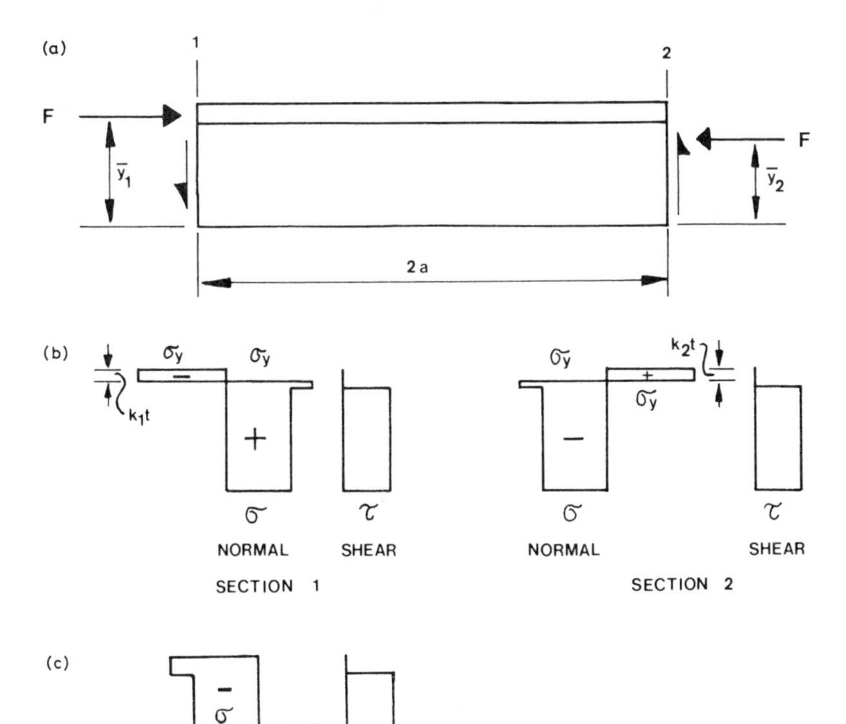

FIG. 4.7. Assumed plastic stress distributions. (a) Tee-section above opening. (b) Stress distributions at sections 1 and 2 for low moment-to-shear ratios. (c) Stress distributions at section 1 for high moment-to-shear ratios.

For the case of high moment-to-shear ratio, eqns (4.7) and (4.8) are replaced by:

$$F = bt\sigma_y + sw\sigma(1 - 2k_1) \tag{4.11}$$

$$F\bar{y}_1 = bt\sigma_y(s + 0{\cdot}5t) + 0{\cdot}5s^2w\sigma(1 - 2k_1^2) \tag{4.12}$$

in which k_1 now has a different definition, as given in Fig. 4.7(c). The normal stress in the web, σ, is related to the shear stress τ (which is assumed constant in the web and zero in the flange) through the yield criterion. Assuming the Von Mises yield criterion, σ is given by

$$\sigma^2 + 3\left(\frac{V}{2sw}\right)^2 = \sigma_y^2 \tag{4.13}$$

In the case of low moment-to-shear ratio, eqns (4.5)–(4.10) and (4.13) can be used to find corresponding values of M and V. If a value of V is assumed, eqn (4.13) gives σ and eqns (4.7) and (4.9) can be used to eliminate k_1 (or k_2). Equation (4.6) then gives k_2 (or k_1) and hence the limit value of the moment M is given by eqn (4.5). If the value of k_1 exceeds unity, the solution for high moment-to-shear ratio which follows the same approach must be used. An explicit presentation of this solution procedure has been given by Aglan and Qaqish (1982).

For a given beam and opening, an interaction curve relating the limit values of M and V can be constructed by the above process. Figure 4.8 shows several diagrams of this type in which the moment M is expressed as a proportion of the plastic moment, M_p, of the unperforated beam section, and the shearing force V is expressed as a fraction of the plastic shear resistance of the unperforated beam, V_p, where

$$M_p = Z\sigma_y = [bt(d-t) + 0{\cdot}25w(d-2t)^2]\sigma_y \tag{4.14}$$

$$V_p = A_w\sigma_y/\sqrt{3} = w(d-2t)\sigma_y/\sqrt{3} \tag{4.15}$$

Results obtained from this method of analysis have been compared with test results obtained by various researchers. The tests embrace widely different values of the ratio of hole height to beam depth, the ratio of hole

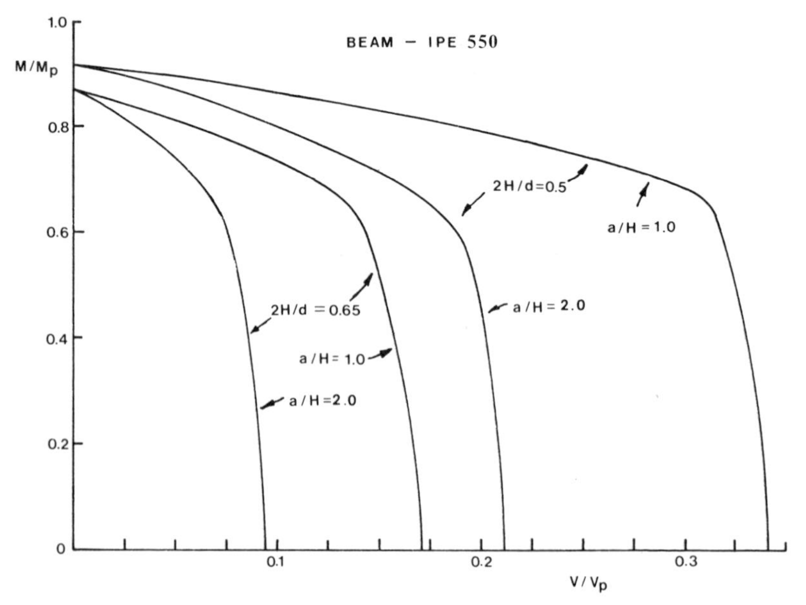

FIG. 4.8. Moment–shear interaction diagrams.

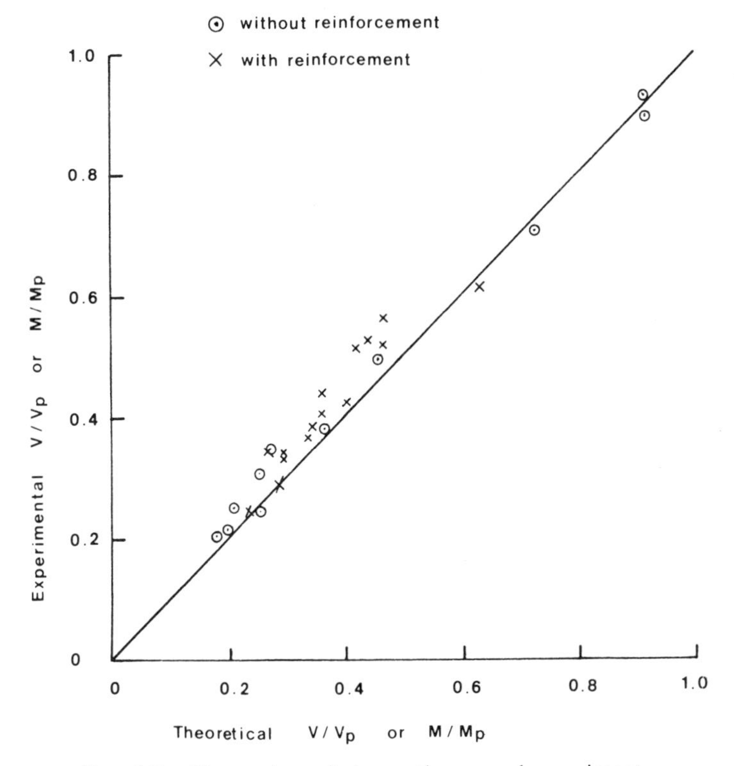

FIG. 4.9. Comparisons between theory and experiment.

length to height and the moment-to-shear ratio, and Fig. 4.9 summarises many of these results. It can be seen that the theory generally gives slightly conservative results, and this is particularly so for low moment-to-shear ratios for which strain-hardening effects can be expected to have an influence. Very close agreement is observed for high moment-to-shear ratios. The results shown in Fig. 4.9 include several cases of circular openings and reinforced openings, the analysis of which will be discussed subsequently.

In the case of low moment-to-shear ratio, eqns (4.5)–(4.10) and (4.13)–(4.15) can be rearranged to give:

$$\left(\frac{V}{V_p}\right)^2 = \left(1 - \frac{2H}{d-2t}\right)^2 - \left(\frac{2bt(k_1 + k_2 - 1)}{w(d-2t)}\right)^2 \tag{4.16}$$

$$\left(\frac{M}{M_p}\right) = \frac{t(k_2^2 - k_1^2) + d(k_1 - k_2)}{(d-t) + w(d-2t)^2/4bt} \tag{4.17}$$

where k_1 and k_2 are related through

$$\left[\left(\frac{sw}{bt}\right)^2 - (k_1+k_2-1)^2\right] = \frac{3}{4a^2}\,[s(k_1+k_2-1) + t\{2k_1k_2 - (k_1+k_2-1)^2\}]^2 \tag{4.18}$$

If it is now assumed that the flange thickness t is small compared with the web height of the tee-section, s, k_2 can be eliminated from these equations, leading to

$$\frac{V}{V_p} = \left(1 - \frac{2H}{d}\right)\sqrt{\left(\frac{\alpha}{1+\alpha}\right)} \tag{4.19}$$

$$\frac{M}{M_p} = \frac{(2k_1 - 1) - \dfrac{A_w}{4A_f}\left(1 - \dfrac{2H}{d}\right)\dfrac{2}{\sqrt{(1+\alpha)}}}{1 + \dfrac{A_w}{4A_f}} \tag{4.20}$$

where A_f = the area of one flange ($= bt$), A_w = the web area ($= w(d-2t)$) and

$$\alpha = \frac{3}{4}\left(\frac{s}{a}\right)^2 = \frac{3}{16}\left(\frac{d}{a}\right)^2\left(1 - \frac{2H}{d}\right)^2 \tag{4.21}$$

This simplification leads to the result that at low moments and high shearing forces, the value of V/V_p remains constant as the moment-to-shear ratio varies and the moment resistance takes on its maximum value, M_1, when $k_1 = 1$, i.e.

$$\frac{M_1}{M_p} = \frac{1 - \dfrac{A_w}{4A_f}\left(1 - \dfrac{2H}{d}\right)\dfrac{2}{\sqrt{(1+\alpha)}}}{1 + \dfrac{A_w}{4A_f}} \tag{4.22}$$

The point on the interaction curve given by eqns (4.19) and (4.22) corresponds to the stress reversal in the top tee at section 1, occurring at the web to flange junction.

In the case of high moment-to-shear ratio, the equations corresponding to (4.19) and (4.20) are as follows:

$$\frac{V}{V_p} = \left(1 - \frac{2H}{d}\right)\sqrt{\left(\frac{\bar{\alpha}}{1+\bar{\alpha}}\right)} \tag{4.23}$$

$$\frac{M}{M_p} = \frac{1 - \dfrac{A_w}{4A_f}\left(1 - \dfrac{2H}{d}\right)\dfrac{1}{\sqrt{(1+\bar{\alpha})}}\left[2k_1(1+0{\cdot}5k_1) - 1 - \dfrac{2H}{d}(1-k_1^2)\right]}{1 + \dfrac{A_w}{4A_f}} \tag{4.24}$$

where

$$\bar{\alpha} = \alpha k_1^2(2 - k_1)^2 \tag{4.25}$$

Thus, for high moment-to-shear ratios, both V and M vary with the parameter k_1. When $k_1 = 1{\cdot}0$, which can be seen from Fig. 4.7(b) to correspond to stress reversal occurring at the junction of web and flange, eqn (4.24) is identical to eqn (4.22). When $k_1 = 0$, the stress distribution corresponds to pure bending, in which case eqn (4.23) gives $V = 0$ and eqn (4.24) reduces to

$$\frac{M_0}{M_p} = 1 - \frac{\dfrac{A_w}{4A_f}\left(\dfrac{2H}{d}\right)^2}{1 + \dfrac{A_w}{4A_f}} \tag{4.26}$$

By varying the parameter k_1 between 0 and 1, the interaction curve for the high moment-to-shear ratio region is obtained from eqns (4.23) and (4.24). Alternatively, the points '0' and '1' at the extremities of this region could more simply be joined by a straight line or an elliptical relationship of the form

$$\left(\frac{V}{V_1}\right)^2 + \left(\frac{M - M_1}{M_0 - M_1}\right)^2 = 1{\cdot}0 \tag{4.27}$$

These three approximations are shown in Fig. 4.10, where they are compared with the more exact curve given by eqns (4.5)–(4.13). The linear approximation is quite conservative, whereas eqn (4.27) is unconservative.

The possibility of yielding in shear on a horizontal plane through the webs of the tee-section has been considered by Shrivastava and Redwood (1979). It is shown that if $2a/s < 3$, such yielding can occur and the interaction diagram should be curtailed at the corresponding value of the shearing force. This value is obtained by first finding k_1 from

$$\sqrt{(3k_1^3[4 - k_1])} = 2a/s \tag{4.28}$$

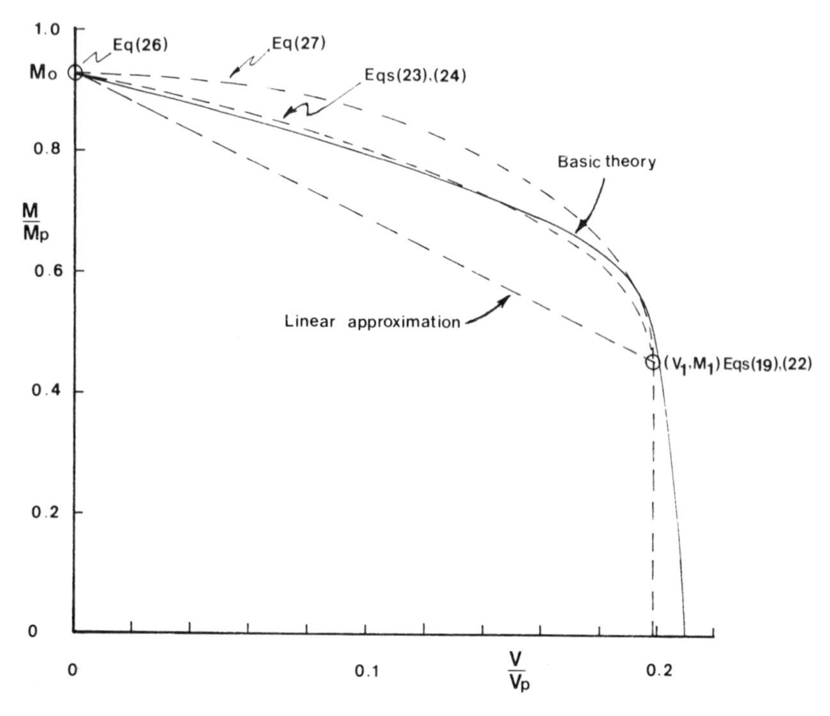

FIG. 4.10. Approximate solutions.

and using this value of k_1 in eqn (4.23). While this will limit the resistance at small openings, its practical significance appears to be largely outweighed by strain-hardening.

The above analysis has been concerned with rectangular openings only. Circular openings can be analysed using the same equations by considering them to be equivalent to rectangular openings with length equal to 0·45 of the hole diameter, and height equal to twice the length. This assumption is based on Redwood (1969), who varied the position of the yield sections and found opening dimensions corresponding to an interaction diagram giving a lower envelope to the corresponding individual diagrams.

(*b*) *Eccentric Openings*

Eccentric unreinforced openings have been studied by Frost (1971) based on stress distributions identical to those shown in Fig. 4.7. The effect of eccentricity is to increase the beam resistance when the moment-to-shear ratio is low and to decrease it when the moment predominates. These effects, particularly the former, are relatively minor.

Simplified interaction diagrams can again be constructed by locating the points corresponding to '0' and '1' on Fig. 4.10. The shear V_1 can be obtained directly by development of eqn (4.19) if it is considered that the maximum shear resistance is given by the sum of the maximum shear resistances of the top and bottom tee-sections. For a mid-depth opening, each tee-section has a resistance V equal to one-half of the beam resistance given by eqn (4.19), i.e.

$$V = \frac{V_p}{2}\left(1 - \frac{2H}{d}\right)\sqrt{\left(\frac{\alpha}{1+\alpha}\right)} = V_p\left(\frac{s}{d}\right)\sqrt{\left(\frac{\alpha}{1+\alpha}\right)} \tag{4.29}$$

where, from Fig. 4.7, $d - 2H = 2s$, and α is given by eqn (4.21).

For the eccentric opening, the tee-sections will have different values of s, as shown on Fig. 4.11, hence the total shear resistance can be written

$$V = V_p\left[\frac{s_1}{d}\sqrt{\left(\frac{\alpha_1}{1+\alpha_1}\right)} + \frac{s_2}{d}\sqrt{\left(\frac{\alpha_2}{1+\alpha_2}\right)}\right] \tag{4.30}$$

or, since $s_1 = 2a\sqrt{(\alpha_1/3)}$ and $s_2 = 2a\sqrt{(\alpha_2/3)}$ from eqn (4.21),

$$\frac{V_1}{V_p} = \frac{2}{\sqrt{3}}\left(\frac{a}{d}\right)\left[\frac{\alpha_1}{\sqrt{(1+\alpha_1)}} + \frac{\alpha_2}{\sqrt{(1+\alpha_2)}}\right] \tag{4.31}$$

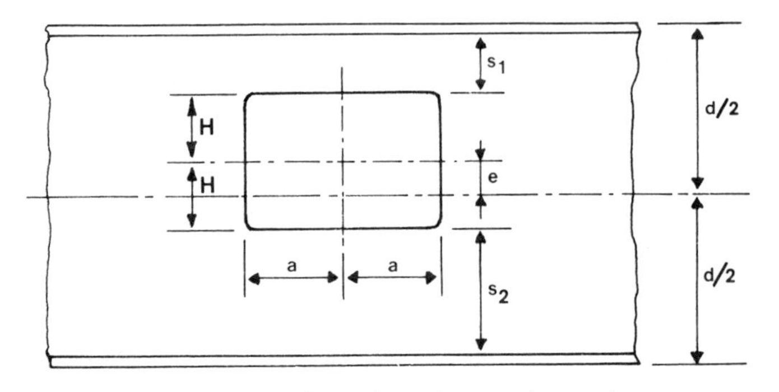

FIG. 4.11. Configuration of eccentric opening.

The parameters α_1 and α_2 may alternatively be written in terms of the eccentricity, e, as

$$\begin{aligned} \alpha_1 &= \frac{3}{16}\left(\frac{d}{a}\right)^2\left[1-\frac{2H}{d}-\frac{2e}{d}\right]^2 \\ \alpha_2 &= \frac{3}{16}\left(\frac{d}{a}\right)^2\left[1-\frac{2H}{d}+\frac{2e}{d}\right]^2 \end{aligned} \tag{4.32}$$

Following the same steps as in the analysis for the mid-depth hole, the following relationships are obtained for bending moment resistances under high shear and pure bending respectively:

$$\frac{M_1}{M_p} = \frac{1-\dfrac{2}{\sqrt{3}}\left(\dfrac{a}{d}\right)\left(\dfrac{A_w}{A_f}\right)\Big/\sqrt{\left(\dfrac{\alpha_2}{1+\alpha_2}\right)}}{1+\dfrac{A_w}{4A_f}} \tag{4.33}$$

$$\frac{M_0}{M_p} = 1 - \frac{\dfrac{A_w}{4A_f}\left[\left(\dfrac{2H}{d}\right)^2+\left(\dfrac{4e}{d}\right)\left(\dfrac{2H}{d}\right)\right]}{1+\dfrac{A_w}{4A_f}} \tag{4.34}$$

The parameter α_2 is always taken as the larger of eqn (4.32) whether the opening centreline is above or below the mid-depth of the beam.

(c) Openings in Composite Beams

Concrete floor slabs and supporting steel beams are frequently connected in order to provide compositely acting beams and girders, and passage of service ducts through the beam webs is often required. Clawson and Darwin (1980, 1982*a*, *b*) have studied this case in detail for solid concrete slabs and have developed a theory, supported by experimental results, to predict the ultimate strength. Two modes of failure are identified, one of which corresponds to the four-hinge mechanism considered above for non-composite beams, and another in which shear failure occurs in the steel tee and concrete slab above the hole at its centreline. For both mechanisms, the tee-section below the opening is fully yielded at both ends in bending, shear and tension. In the first failure mode, cracking of concrete is assumed to occur at the low moment end of the opening, and in both modes, concrete failure corresponds to a failure criterion involving shearing and normal stresses. The effects of longitudinal slab reinforcement are included in the model. A comparison between this theory and test results is shown in Fig. 4.12(a).

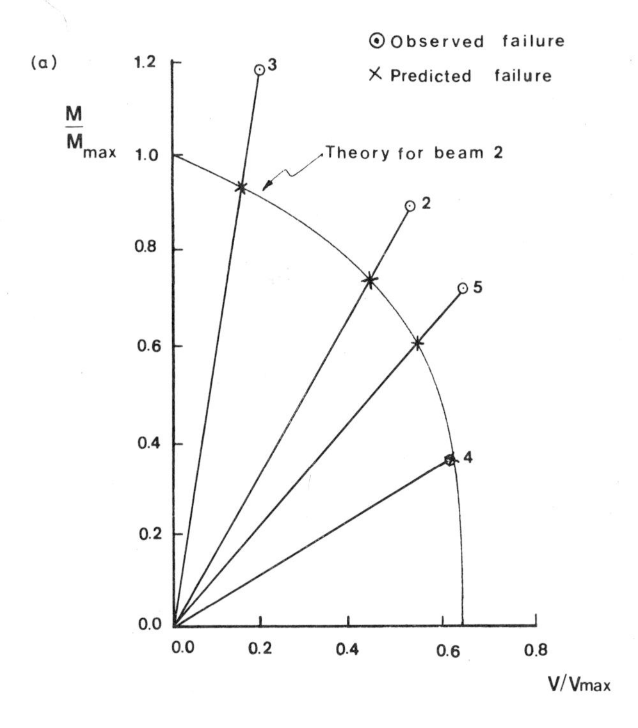

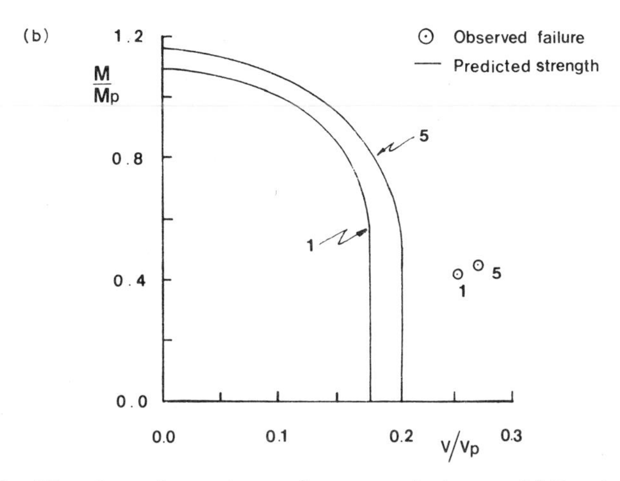

FIG. 4.12. Theories and experiments for composite beams. (a) Results for solid slab beams (from Clawson and Darwin, 1982*b*). (b) Results for ribbed deck supported slabs (from Redwood and Wong, 1982).

Redwood and Wong (1982) have considered steel deck supported slabs composite with the steel beam, and have treated the shear carried by the concrete quite conservatively. The analysis is of a four-hinge mechanism and follows closely the approximate approach outlined above for non-composite beams. This analysis will now be described and is applied to an opening with eccentricity, e, above the mid-depth of the beam as shown in Fig. 4.13(a).

An interaction curve of the same form as that shown in Fig. 4.10 can be constructed. The pure bending moment resistance at the opening, M_0, can be found using standard ultimate strength calculations for composite beams. Because an opening may be close to the end of a beam, this bending resistance will frequently be influenced by the limited amount of shear

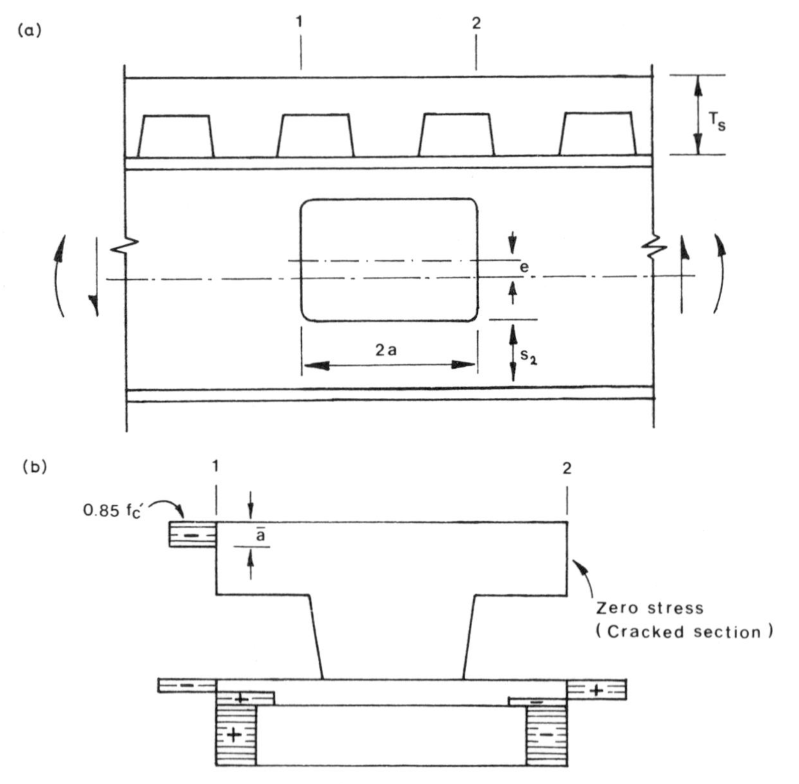

FIG. 4.13. Assumed ultimate stress distributions: composite beam. (a) Eccentric opening in a composite beam. (b) Stress distributions above opening when maximum shear force is acting.

connection provided in the length between the opening centreline and the nearest point of zero moment. The maximum shear resistance can be obtained as the sum of two parts, the resistance of the tee-section below the opening and that of the slab and steel tee above the opening. The former, V_{b1}, following eqn (4.31) can be written

$$V_{b1} = V_p \frac{2}{\sqrt{3}} \left(\frac{a}{d}\right) \frac{\alpha_2}{\sqrt{(1+\alpha_2)}} \tag{4.35}$$

To evaluate the maximum shear resistance of the section above the opening, full yielding at each end is assumed with the stresses distributed as shown in Fig. 4.13(b). These distributions correspond to cracking of the concrete slab over its full depth at the low moment end, and to a compression block in the concrete at the high moment end in which the total compressive force is equal to the ultimate capacity of the shear connectors provided within the length of the opening. Thus, the depth of compression block, $\bar{a}$, can be written

$$\bar{a} = \frac{n_h q}{0{\cdot}85 f_c' b_e} \tag{4.36}$$

in which n_h is the number of shear connectors within the length of the opening, q is the ultimate resistance per connector, b_e is the effective width of the slab and f_c' is the compression (cylinder) strength of concrete.

If it is again assumed that the flange thickness t is small compared with the depth of the steel tee-sections s_1 and s_2, equilibrium of the upper tee-section leads to the maximum shear resistance, V_{t1}, given by

$$V_{t1}^2(3+\gamma^2) - 2\mu\gamma V_{pt} V_{t1} + (\mu^2 - 3)V_{pt}^2 = 0 \tag{4.37}$$

where

$$\gamma = 2a/s_1 = \sqrt{(3/\alpha_1)} \tag{4.38}$$

$$V_{pt} = s_1 w F_y/\sqrt{3} \tag{4.39}$$

$$\mu = \frac{n_h q}{V_{pt}} \left[\frac{T_s - 0{\cdot}5\bar{a}}{s_1}\right] \tag{4.40}$$

and T_s is the overall slab depth. The maximum possible value of V_{t1} is V_{pt}, and this can be attained if $\mu \geq \gamma$. The maximum shear strength of the beam at the opening can now be written

$$V_1 = V_{b1} + V_{t1} \tag{4.41}$$

and it can also be shown, following the same steps which led to eqn (4.33), that while supporting this maximum shear force the beam has a bending resistance given by:

$$\frac{M_1}{M_p} = \frac{1 - \frac{2}{\sqrt{3}}\left(\frac{a}{d}\right)\left(\frac{A_w}{A_f}\right)\left[\sqrt{\left(\frac{\alpha_2}{1+\alpha_2}\right)} - \frac{\mu\alpha_1}{3}\left(\frac{a}{d}\right)\right]}{1 + \frac{A_w}{4A_f}} \tag{4.42}$$

where, as before, M_p is the plastic moment of the unperforated steel beam section.

With these values of V_1, M_1 and M_0, an interaction diagram can be constructed using either eqn (4.27) or a linear approximation between points '0' and '1'. Two such diagrams are shown in Fig. 4.12(b) and are compared with test results.

This method has the advantage of simplicity, but is conservative partly because it takes into account the limited shear connection over the length of the opening while at the same time ignoring compression in the slab at the low moment end of the opening which would make the limited shear connection less critical. This compression has been observed in solid slab composite beams by Clawson and Darwin (1980), who also exclude it from their analytical model but at the same time do not include the restriction of the shear connection provided within the opening length. In other respects this model is quite comprehensive, and while its complexity prevents its use as a design tool, it can be used to verify proposed design rules. Further studies of composite beams are needed with the aim of improving these analytical approaches and developing design aids.

4.3.3 Openings with Reinforcement

Reinforcement of openings by means of horizontal bars such as shown in Fig. 4.1(a) and (b) is a relatively economical way of increasing the resistance at an opening. Typically the bars are welded on both sides of the web to preserve symmetry, and they are slightly offset from the opening edge to facilitate welding. Because of its economy, this type of reinforcement has been studied in some detail.

Congdon and Redwood (1970) analysed beams with mid-depth openings using the same stress distributions summarised in Fig. 4.7, with the added assumption that the reinforcing bars carried no shearing force directly. The most general case, that of reinforced eccentric openings, was treated by Wang *et al.* (1975), based on similar assumed stress distributions. Six different solutions have to be considered according to whether the stress reversals at the different hinge locations occur in the flange, in the web between flange and reinforcement, within the depth of reinforcement or in the web stub between the reinforcement and the opening. These analyses have been compared with test results, and agreement is as good as in the unreinforced cases as shown in Fig. 4.9.

By assuming that flanges and reinforcing bars are thin compared with the tee-web depth, and by assuming that the reinforcement is placed at the edges of the opening, simplified solutions were obtained and are summarised by Kussman and Cooper (1976). The coordinates of the points '1' and '0' on the interaction diagram can be written as follows.

For $A_r < A_{r0}$:

$$\left(\frac{V_{b1}}{V_p}\right)^2 = \frac{4}{3}\left(\frac{a}{d}\right)^2 \alpha_2 - \left(\frac{A_f}{A_w}\beta_2\right)^2 \quad (4.43)$$

$$\left(\frac{V_{t1}}{V_p}\right)^2 = \frac{4}{3}\left(\frac{a}{d}\right)^2 \alpha_1 - \left(\frac{A_f}{A_w}\beta_1\right)^2 \quad (4.44)$$

and the total shear resistance corresponding to point '1' is given by eqn (4.40).

$$\frac{M_1}{M_p} = \frac{1 - \dfrac{A_r}{A_f} - \beta_2}{1 + \dfrac{A_w}{4A_f}} \quad (4.45)$$

For $A_r \geq A_{r0}$:

$$\frac{V_1}{V_p} = 1 - \frac{2H}{d} \quad (4.46)$$

$$\frac{M_1}{M_p} = \frac{1 - \dfrac{A_{r0}}{A_f}}{1 + \dfrac{A_w}{4A_f}} \quad (4.47)$$

In the above:

$$A_{r0} = aw/\sqrt{3} \quad (4.48)$$

$$\beta_1 = -\frac{2\alpha_1}{1+\alpha_1}\left(\frac{A_r}{A_f}\right) + 2\left(\frac{A_w}{A_f}\right)\sqrt{\left(\frac{\alpha_1}{1+\alpha_1}\right)}\left[\frac{1}{\sqrt{3}}\left(\frac{a}{d}\right)^2 - \frac{1}{1+\alpha_1}\left(\frac{A_r}{A_w}\right)\right]^{1/2} \quad (4.49)$$

$$\beta_2 = -\frac{2\alpha_2}{1+\alpha_2}\left(\frac{A_r}{A_f}\right) + 2\left(\frac{A_w}{A_f}\right)\sqrt{\left(\frac{\alpha_2}{1+\alpha_2}\right)}\left[\frac{1}{\sqrt{3}}\left(\frac{a}{d}\right)^2 - \frac{1}{1+\alpha_2}\left(\frac{A_r}{A_w}\right)\right]^{1/2} \quad (4.50)$$

The shear given by eqn (4.46) is the maximum which can be carried by the available web area. If the applied factored shear force exceeds this value, then this type of reinforcement is not suitable. The pure bending resistance is reduced as a result of eccentricity and can be written (see Redwood, 1978*a*):

$$\left(\frac{M_0}{M_p}\right)' = 1 + \frac{\dfrac{A_r}{A_f}\left(\dfrac{2H}{d}\right) - \dfrac{A_w}{4A_f}\left[\left(\dfrac{2H}{d}\right)^2 + 4\left(\dfrac{2H}{d}\right)\left(\dfrac{2e}{d}\right) - 4\left(\dfrac{2e}{d}\right)^2\right]}{1 + \dfrac{A_w}{4A_f}}$$

$$\text{for} \quad \left|\frac{e}{d}\right| \leq \frac{A_r}{A_w} \qquad (4.51)$$

$$\left(\frac{M_0}{M_p}\right)'' = \left(\frac{M_0}{M_p}\right)' - \frac{\dfrac{A_w}{A_f}\left[\dfrac{e}{d} - \dfrac{A_r}{A_w}\right]^2}{1 + \dfrac{A_w}{4A_f}} \qquad \text{for} \quad \left|\frac{e}{d}\right| > \frac{A_r}{A_w} \qquad (4.52)$$

If the area of reinforcement is relatively small, it has been shown by Lupien and Redwood (1978) that horizontal bar reinforcement can, under certain conditions, be placed on one side only of the web as shown in Fig. 4.1(b). This is possible where strain-hardening compensates for the reduction in resistance due to lateral bending. This arrangement leads to economies in both welding and handling of the beam. The necessary conditions for the above theoretical analyses to be used directly for one-sided reinforcement were determined experimentally for mid-depth openings and are as follows:

$$\begin{aligned} &A_r \leq 0{\cdot}333A_f \\ &a/H \leq 2{\cdot}5 \\ &\frac{d-2t}{w}\sqrt{\left(\frac{\sigma_y}{E}\right)} \leq 2{\cdot}5 \\ &\frac{b_r}{t_r}\sqrt{\left(\frac{\sigma_y}{E}\right)} \leq 0{\cdot}32 \\ &\frac{s}{w}\sqrt{\left(\frac{\sigma_y}{E}\right)} \leq 0{\cdot}82 \end{aligned} \qquad (4.53)$$

and $M/Vd \leq 20$ at the opening centreline. In the above, b_r and t_r are the width and thickness of the reinforcing bars, $A_r = b_r t_r$, and s is the larger of s_1 or s_2.

A number of alternative types of reinforcement have been studied by Segner (1964). These include types (c), (d), (e), (g) and (h) shown in Fig. 4.1. While methods of analysis are not available for most of these, the observed test behaviour and measured strengths give useful insight into their performance. In view of its economy and the availability of reliable methods of analysis, the horizontal bar type of reinforcement appears to have clear advantages over most other types, except under conditions of very severe shear loading. In selection of a reinforcement type, the structural actions in the region around the opening should be carefully considered. Owing to shear-induced flexure in the regions above and below the opening, horizontal reinforcement close to the top and bottom edges of the opening will carry high axial loads which will peak at the opening corners. Adequate anchorage for such forces can be developed by providing extensions beyond the corners, for example in types (a), (b) and (c) in Fig. 4.1. Types (d) and (e) appear to be poor configurations for this purpose, since very high loading will be produced on the web and welds at the corners. When shear forces are very high and the maximum available web shear resistance given by eqn (4.46) is exceeded, direct means of enhancing the web shear-carrying capacity by using types (g) to (j) must be considered. Except for this case, it appears that the horizontal bar type of reinforcement has clear advantages over many other types because of its simplicity, hence economy, and the availability of reliable methods of analysis.

4.3.4 Multiple Openings

A beam in the region of adjacent openings may fail in a mode in which interaction between the openings takes place if the spacing between openings is small. In addition, the shearing force will have to be sufficiently high, since under pure bending conditions no interaction will occur. The interaction results from deformation of the web-post between the openings. This post is bent in double curvature and may undergo plastic hinging at its ends, or it may buckle laterally. Furthermore, if the post varies in width (as in the case of circular openings or in castellated beams), it may yield in shear at its narrowest section.

In the case where a plastic mechanism forms, as shown in Fig. 4.14(a), the limit loads for the single opening mode can be obtained using eqns (4.5)–(4.15), and a similar analysis can be carried out for the interacting mode (Redwood, 1968*a*). These diagrams, shown in Fig. 4.14(b), illustrate the dependence of the interaction on the moment-to-shear ratio. The analysis of the interacting mode, taking into account full interaction of bending moment, axial force and shearing force at the various hinge locations, is laborious if the procedure follows that described above for the single opening. Alternatively, a simple approximate solution can be obtained in which the shear which will cause the interacting collapse mode is given by

$$V = \frac{4M_{pT} + 2\lambda M_{post}}{g + 2a} \tag{4.54}$$

in which M_{pT} = plastic moment of the tee-section above or below the opening, M_{post} = plastic moment of the web-post at the section where failure occurs, g = centre-to-centre distance between openings, and λ = a geometrical parameter which depends upon the opening geometry.

For rectangular openings, $\lambda = 1 + \bar{y}/H$, where $\bar{y}$ is the distance from the edge of the opening to the plastic neutral axis of the tee-section, and the plastic hinges form at the top and bottom of the post. λ and the hinge locations for other opening shapes are given in the reference. While this equation can be solved iteratively by adjusting the values of the plastic moments M_{pT} and M_{post} at each step to reflect the axial and shearing forces acting at each yielding cross-section, it can be noted that the single opening solution limits the application of the double opening solution to regions of low moment-to-shear ratio. Thus, the normal force effects on the plastic moments will be small, and it is found that eqn (4.54) based upon the unadjusted values of M_{pT} and M_{post} gives a shear cut-off on the interaction diagram which is in reasonable agreement with the lower bound solution. An example is shown in Fig. 4.14(b), and the procedure is also shown to be safe when compared with available test results (Redwood, 1968*b*). A convenient way to increase the shear resistance of the interacting mode, to avoid interaction if necessary, is to increase the plastic moment of the web-post by attaching vertical reinforcing bars at the edges of the post.

Buckling of web-posts has been studied by Aglan and Redwood (1974) and Dougherty (1981), who show that such buckling will not in general occur in rolled sections prior to development of a yield mechanism, and the latter reference indicates that such buckling is possible in plate girders. Redwood and Shrivastava (1980) give lower limits of the web-post width of $2H$ for rectangular and $3R$ for circular openings such that neither buckling nor an interacting collapse mechanism will occur.

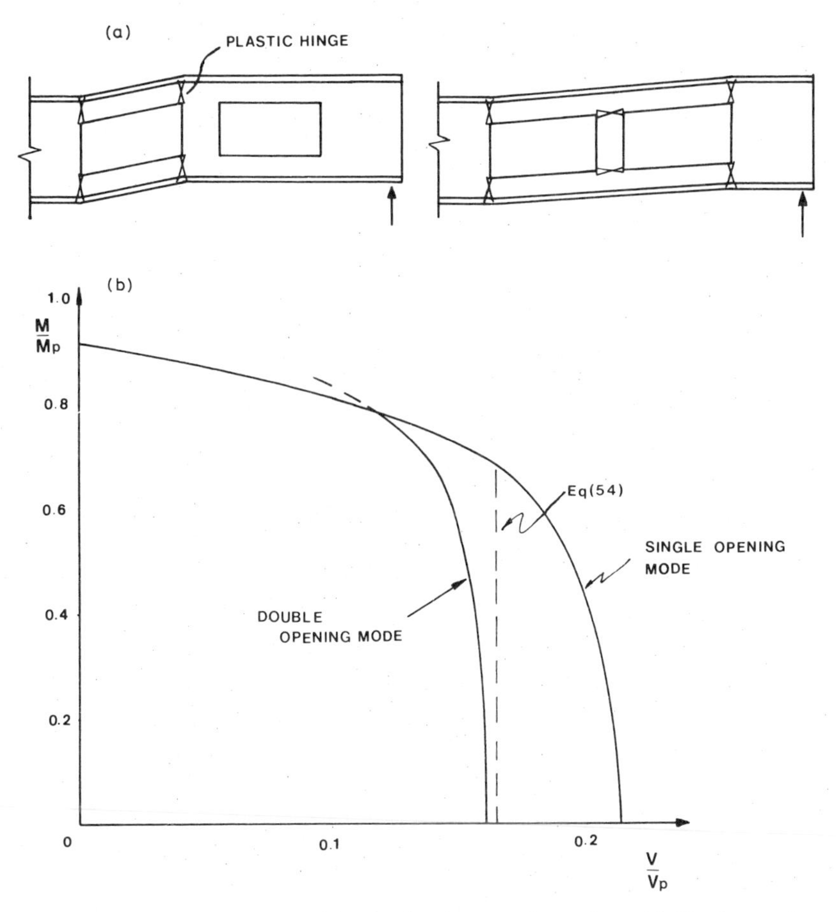

FIG. 4.14. Failure of beam with two closely spaced openings. (a) Single opening and interacting failure modes. (b) Interaction diagrams.

4.3.5 Buckling

The principal need for large web openings arises in building structures, and consequently most of the results presented relate to beam sections typical of floor framing systems. Such beams are usually provided with closely spaced lateral supports, and have web and flange width-to-thickness ratios such that the yield stress can be reached prior to the occurrence of local buckling. For these beams, therefore, there is little concern with buckling problems, with the exception of those treated by means of plastic analysis for which rotation capacity must be available for the collapse mechanisms to develop.

Specific recommendations have been presented by Redwood and Uenoya (1979) which give limits on the opening and beam parameters within which local buckling need not be considered. For example, for beams with flange slenderness $b/2t \leq 0{\cdot}380\sqrt{(E/\sigma_y)}$ and web slenderness $(d-2t)/w \leq 3{\cdot}05\sqrt{(E/\sigma_y)}$, mid-depth opening can be analysed according to the ultimate strength methods described in Section 4.3.2, provided the shearing force applied at the opening is less than $0{\cdot}45V_p$, and if

$$\frac{a}{H} \leq 2{\cdot}2 \qquad \text{and} \qquad \frac{a}{H} + 6\left(\frac{2H}{d}\right) \leq 5{\cdot}6 \tag{4.55}$$

These limits are relaxed for webs with stockier sections. In addition, Shrivastava and Redwood (1977) recommend that for reinforced openings in webs satisfying the above slenderness limit, no concern with web buckling arises if the calculated area of reinforcement required does not exceed A_f or $0{\cdot}5A_w$, whichever is less. Höglund and Johansson (1977) provide graphical information indicating that no reduction in strength occurs due to buckling if the web slenderness limit is about $2{\cdot}4\sqrt{(E/\sigma_y)}$.

These results all relate to mid-depth openings. Eccentric openings may produce more critical conditions in view of the slender tee-section web on one side of the opening. Little information is available related to this case.

Beams with very slender webs are beyond the scope of this chapter. It has been shown by Höglund (1971) and Narayanan and Rockey (1981) that significant post-buckling strength exists in such beams, and methods of predicting the ultimate capacity have been presented.

4.4 DESIGN INFORMATION

Several of the solutions described in the preceding sections have been used as a basis for design aids. Recommendations by Bower *et al.* (1971) and Redwood (1973) deal with mid-depth unreinforced openings; they treat allowable stress design by Vierendeel analysis ignoring stress concentrations, and treat ultimate strength design using eqns (4.19), (4.22) and (4.26). Interaction diagrams based on theory of elasticity and approximate elastic analysis are given by Bower (1968*a*) and Redwood and Chan (1974). Höglund and Johansson (1977) give comprehensive recommendations and graphical design aids dealing with thin-webbed beams, as well as the stockier beams considered herein. For the latter, the ultimate strength equations used are similar to those quoted above. Redwood (1972, 1978*b*) gives, in tabulated form, explicit values of required reinforcement areas for mid-depth openings and eccentric openings respectively for any given loading. The latter are obtained, after further simplifying assumptions, from eqns (4.43)–(4.52). Redwood and Shrivastava (1980) give comprehensive design recommendations based on plastic analysis. Recent design aids incorporating tabulated data have been provided by United States Steel (1981) and Canadian Institute of Steel Construction (1980). Design details have been treated by Höglund and Johansson (1977) and Redwood and Shrivastava (1980). These details include anchorage requirements for reinforcement, effects of concentrated loads and welding details.

It is evident that there is a wealth of information available which will permit satisfactory approaches for the design of beams with web openings for allowable stress and ultimate strength design. Only one type of reinforcement has been studied in detail, but this type will be suitable for many practical situations. However, there are some applications of practical importance about which further information is needed. Foremost among these are composite beams, including the effects of unshored construction on the resistance, and the effects on the concrete slab of local deflections at an opening. Development of design rules for these beams which are less conservative than those described in Section 4.3.2(c) is also desirable. Other areas in which more information would be useful include the effects of eccentric openings on web stability, and the effects of concentrated loads near openings.

REFERENCES

AGLAN, A. A. and QAQISH, S. (1982) Plastic behaviour of beams with mid-depth web openings. *AISC Engineering Journal*, **19**(1), 20–6.

AGLAN, A. A. and REDWOOD, R. G. (1974) Web buckling in castellated beams. *Proc. Inst. Civ. Engrs.*, **57**, 307–20.

BOWER, J. E. (1966*a*) Elastic stresses around holes in wide-flange beams. *Journal of the Structural Division, Proc. ASCE*, **92**(ST2), 85–101.

BOWER, J. E. (1966*b*) Experimental stresses in wide flange beams with holes. *Journal of the Structural Division, Proc. ASCE*, **92**(ST5), 167–86.

BOWER, J. E. (1967) Ultimate strength of wide-flange beams with rectangular holes. United States Steel Corporation, Applied Research Laboratory Report 57.019-400(13) (Feb.).

BOWER, J. E. (1968*a*) Design of beams with web openings. *Journal of the Structural Division, Proc. ASCE*, **94**(ST3), 783–807. See also US Steel, Building Design Data, ADUSS 27-3500-01 (Apr. 1968) and preprint no. 493, ASCE Structural Engineering Conference (May 1967).

Bower, J. E. (1968*b*) Ultimate strength of beams with rectangular holes. *Journal of the Structural Division, Proc. ASCE*, **94**(ST6), 1315–37.

Bower, J. E. (1971) Recommended design procedures for beams with web openings. *AISC Engineering Journal*, **8**(4), 132–7.

Bower, J. E. *et al.* (1971) Suggested design guides for beams with web holes. *Journal of the Structural Division, Proc. ASCE*, **97**(ST11), 2707–28.

Brock, J. S. (1958) The stresses around square holes with rounded corners. *Journal of Ship Research*, **2**(2), 37–41.

Canadian Institute of Steel Construction (1980) *Handbook of Steel Construction*, 3rd Edn, CISC, Toronto.

Chan, P. (1971) Approximate methods to calculate stresses around circular holes. Fourth Progress Report to Canadian Steel Industries Construction Council, Project 695 (Nov.).

Chan, P. W. and Redwood, R. G. (1974) Stresses in beams with circular eccentric web holes. *Journal of the Structural Division, Proc. ASCE*, **100**(ST1), 231–48.

Clawson, W. C. and Darwin, D. (1980) Composite beams with web openings. Report on NSF Grant No. ENG 76-19045, Structural Engineering and Engineering Materials, SM Report No. 4 (Oct.), University of Kansas Center for Research, Lawrence, Kans.

Clawson, W. C. and Darwin, D. (1982*a*) Tests of composite beams with web openings. *Journal of the Structural Division, Proc. ASCE*, **108**(ST1), 145–62. 1), 145–62.

Clawson, W. C. and Darwin, D. (1982*b*) Strength of composite beams at web openings. *Journal of the Structural Division, Proc. ASCE*, **108**(ST3), 623–41.

Congdon, J. G. and Redwood, R. G. (1970) Plastic behaviour of beams with reinforced holes. *Journal of the Structural Division, Proc. ASCE*, **96**(ST9), 1933–56.

Cooper, P. B. and Snell, R. R. (1972) Tests on beams with reinforced web openings. *Journal of the Structural Division, Proc. ASCE*, **98**(ST3), 611–32.

Cooper, P. B., Snell, R. R. and Knostman, H. D. (1977) Failure tests on beams with eccentric web holes. *Journal of the Structural Division, Proc. ASCE*, **103**(ST9), 1731–8.

Deresiewicz, H. (1968) Stresses in beams having holes of arbitrary shape. *Journal of the Engineering Mechanics Division, Proc. ASCE*, **94**(EM5), 1183–214.

Dougherty, B. K. (1980) Elastic deformation of beams with web openings. *Journal of the Structural Division, Proc. ASCE*, **106**(ST1), 301–12.

Dougherty, B. K. (1981) Buckling of web posts in perforated beams. *Journal of the Structural Division, Proc. ASCE*, **107**(ST3), 507–19.

Douglas, T. R. and Gambrell, Jr, S. C. (1974) Design of beams with off-center web openings. *Journal of the Structural Division, Proc. ASCE*, **100**(ST6), 1189–203.

Frost, R. W. (1971) The behaviour of steel beams with eccentric web holes. Paper presented at ASCE Conference, St. Louis, Mo. (Oct.).

Frost, R. W. and Leffler, R. E. (1971) Fatigue tests of beams with rectangular web holes. *Journal of the Structural Division, Proc. ASCE*, **97**(ST2), 509–27.

Gibson, J. E. and Jenkins, W. M. (1956) The stress distribution in a simply-supported beam with a circular hole. *Structural Engineer*, **34**(12), 443–9.

GOTOH, K. (1975*a*) The stresses in wide-flange beams with web holes. *Theoretical and Applied Mechanics*, **23**, 223–42 (University of Tokyo Press).

GOTOH, K. (1975*b*) Stress analysis of castellated beams. *Trans. Japan Soc. Civil Engineers*, **7**, 37–8.

HALLEUX, P. (1967) Limit analysis of castellated steel beams. *Acier–Stahl–Steel*, No. 3, 133–44.

HELLER, Jr, S. R. (1951) The stresses around a small opening in a beam subjected to bending with shear, *Proc. 1st US National Congress of Applied Mechanics*, ASME, Chicago, pp. 239–45.

HELLER, Jr, S. R. (1953) Reinforced circular holes in bending with shear. *Journal of Applied Mechanics*, **20**, 279–85.

HELLER, Jr, S. R., BROCK, J. S. and BART, R. (1958) The stresses around a rectangular opening with rounded corners in a uniformly loaded plate. *Proc. 3rd US National Congress of Applied Mechanics*, ASME, pp. 357–68.

HELLER, Jr, S. R., BROCK, J. S. and BART, R. (1962) The stresses around a rectangular opening with rounded corners in a beam subjected to bending with shear. *Proc. 4th US National Congress of Applied Mechanics*, ASME, Vol. 1, pp. 489–96.

HÖGLUND, T. (1971) Strength of thin plate girders with circular or rectangular web holes without web stiffeners. *Reports of the Working Commissions*, IABSE Colloquium, London, Vol. 11, pp. 353–65.

HÖGLUND, T. and JOHANSSON, B. (1977) Stegbleche mit Öffnungen. Chap. 5 in *Steifenlose Stahlskeletttragwerke und dünnwandige Vollwandträger*, ed. F. Reinitzhuber, Verlag von Wilhelm Ernst, Berlin, pp. 100–12.

HOPE, B. B. and SHEIKH, M. A. (1969) The design of castellated beams. *Transactions, EIC*, EIC-69-BR and STR 6, **12**(A-8), 1–9.

HOSAIN, M. U. and SPEIRS, W. G. (1973) Experiments on castellated beams. Welding Research Supplement, *Welding Journal*, **52**(8), 329–342-s.

KUSSMAN, R. L. and COOPER, P. B. (1976) Design example for beams with web openings. *AISC Engineering Journal*, **13**(2), 48–56.

LUPIEN, R. and REDWOOD, R. G. (1978) Steel beams with web openings reinforced on one side. *Canadian Journal of Civil Engineering*, **5**(4), 451–61.

MCCORMICK, M. M. (1972) Open web beams: behaviour analysis and design. BHP Melbourne Research Laboratories, Report 17/18, Melbourne, Australia.

MCCUTCHEON, J. O., SO, W.-C. and GERSOVITZ, B. (1963) A study of the effects of large circular openings in the webs of wide flange beams. Applied Mechanics Series No. 2, McGill University, Montreal (Nov.).

MANDEL, J. A., BRENNAN, P. J., WASIL, B. A. and ANTONI, C. M. (1971) Stress distribution in castellated beams. *Journal of the Structural Division, Proc. ASCE*, **97**(ST7), 1947–67.

NARAYANAN, R. and ROCKEY, K. C. (1981) Ultimate load capacity of plate girders with webs containing circular cut-outs. *Proc. Inst. Civ. Engrs.*, **71**(2), 845–62.

OLANDER, H. C. (1953) A method of calculating stresses in rigid frame corners. *Journal of the Structural Division, Proc. ASCE*, **79**, Separate No. 249, 1–21.

REDWOOD, R. G. (1968*a*) Plastic behaviour and design of beams with web openings. *Proc. 1st Canadian Structural Engineering Conference*, Canadian Steel Industries Construction Council, Toronto (Feb.), pp. 127–38.

REDWOOD, R. G. (1968*b*) Ultimate strength design of beams with multiple openings. Preprint No. 757, ASCE Structural Engineering Conference, Pittsburg (Oct.).

REDWOOD, R. G. (1969) The strength of steel beams with unreinforced web holes. *Civil Engineering and Public Works Review (London)*, **64**(755), 559–62.

REDWOOD, R. G. (1971) Stresses in webs with circular openings. Final Report to the Canadian Steel Industries Construction Council, Research Project No. 695 (Dec.).

REDWOOD, R. G. (1972) Tables for plastic design of beams with rectangular holes. *AISC Engineering Journal*, **9**(1), 2–19.

REDWOOD, R. G. (1973) *Design of Beams with Web Holes*. Canadian Steel Industries Construction Council.

REDWOOD, R. G. (1978*a*) Analyse et dimensionnement des poutres ayant des ouvertures dans les âmes. *Construction Métallique*, No. 3, 15–27.

REDWOOD, R. G. (1978*b*) Dimensionnement du renfort d'âme de poutres comportant une ouverture. CRIF (Centre de Recherches Scientifiques et Techniques de l'Industrie des Fabrications Métalliques), Bruxelles, Note Technique No. 16.

REDWOOD, R. G. and CHAN, P. W. (1974) Design aids for beams with circular eccentric web holes. *Journal of the Structural Division, Proc. ASCE*, **100**(ST2), 297–303.

REDWOOD, R. G. and SHRIVASTAVA, S. C. (1980) Design recommendations for steel beams with web holes. *Canadian Journal of Civil Engineering*, **7**(4), 642–50.

REDWOOD, R. G. and UENOYA, M. (1979) Critical loads for webs with holes. *Journal of the Structural Division, Proc. ASCE*, **105**(ST10), 2053–67.

REDWOOD, R. G. and WONG, P. K. (1982) Web holes in composite beams with steel deck. *Proc. Canadian Structural Engineering Conference*, Canadian Steel Construction Council, Vancouver (Feb.).

ROCKEY, K. C., ANDERSON, R. G. and CHEUNG, Y. K. (1969) The behaviour of square shear webs having a circular hole. In *Thin Walled Steel Structures*, ed. K. C. Rockey and H. V. Hill, Crosby Lockwood, London, pp. 148–72.

SAHMEL, P. (1969) Konstruktive Ausbildung und Näherungsberechnung geschweisster Biegeträger und Torsionsstäbe mit grossen Stegausnehmungen [The design, construction and approximate analysis of welded beams and torsion members having large web openings]. *Schweissen und Schneiden*, **21**(3), 116–22.

SAVIN, G. N. (1961) *Stress Concentration Around Holes*. Pergamon Press, Oxford.

SEELY, F. B. and SMITH, J. O. (1967) *Advanced Mechanics of Materials*, Wiley, New York.

SEGNER, Jr, E. P. (1964) Reinforcement requirements for girder web openings. *Journal of the Structural Division, Proc. ASCE*, **90**(ST3), 147–64.

SHERBOURNE, A. N. and VAN OOSTROM, J. (1972) Plastic analysis of castellated beams. I. Interaction of moment, shear and axial force. *Computers and Structures*, **2**, 79–109.

SHRIVASTAVA, S. C. and REDWOOD, R. G. (1977) Web instability near reinforced rectangular holes. *IABSE Proceedings*, No. P-6 (Aug.).

SHRIVASTAVA, S. C. and REDWOOD, R. G. (1979) Shear carried by flanges at unreinforced web holes. Technical Note, *Journal of the Structural Division, Proc. ASCE*, **105**(ST8), 1706–11.

UENOYA, M. and OHMURA, H. (1972) Finite element method for elastic–plastic analysis of beams with holes. Presented at Japan Soc. Civil Engrs. National Meeting, Fukuoka (Oct.).

UNITED STATES STEEL (1981) Web penetrations in steel beams. Design Aid, ADUSS 21-7108-01 (June).

VAN OOSTROM, J. and SHERBOURNE, A. N. (1972) Plastic analysis of castellated beams. II. Analysis and tests. *Computers and Structures*, **2**, 111–40.

WANG, C.-K. (1946) Theoretical analysis of perforated shear webs. *Trans. ASME, Journal of Applied Mechanics*, **13**, A-77–84.

WANG, C.-K., THOMAN, W. H. and HUTCHINSON, C. A. (1955) Stresses in shear webs contiguous to large holes. Internal report, University of Colorado.

WANG, T.-M., SNELL, R. R. and COOPER, P. B. (1975) Strength of beams with eccentric reinforced holes. *Journal of the Structural Division, Proc. ASCE*, **101**(ST9), 1783–1800.

WORLEY, W. J. (1958) Inelastic behavior of aluminum alloy I-beams with web cutouts. Bulletin No. 448, University of Illinois Engineering Experiment Station, Urbana, Ill. (Apr.).

Chapter 5

INSTABILITY, GEOMETRIC NON-LINEARITY AND COLLAPSE OF THIN-WALLED BEAMS

T. M. ROBERTS

Department of Civil and Structural Engineering,
University College, Cardiff, UK

SUMMARY

Non-linear expressions for the strains occurring in thin-walled bars of open cross-section when subjected to axial, flexural and torsional displacements are used to investigate the instability, geometric non-linearity and collapse of such structures, using energy methods. Although the examples presented deal mainly with beams, the term 'bar' is used throughout to imply that the analysis is applicable for all loading and support conditions. It is shown that the influence of pre-buckling displacements is automatically included in the analysis of instability, and closed form solutions obtained agree exactly with solutions based on the governing non-linear differential equations. The geometrically non-linear analysis is combined with an approximate failure criterion to predict lateral failure of simply supported beams and the results thus obtained are shown to agree favourably with experimental results.

NOTATION

a	Length of beam or column
b	Width of flange
d	Depth of beam
l_e	Effective length
q_i	Generalised displacement
r	Radius of gyration of cross-section

r_i, r_j	Generalised displacements of external forces
u, v, w	Displacements in x, y and z directions
v_c	Value of v at centre of beam or column
v_0, w_0	Initial displacements (imperfections) in the y and z directions
v_{0c}	Value of v_0 at the centre of a beam or column
x, y, z	Coordinate directions: distance from centroid in x, y and z directions
A	Cross-section area
C_w	Warping constant
C_y, C_z, C_ρ, $C_{\rho w}$	Section properties
C_1, C_2, C_3	Maximum amplitude of displacement variations
E	Young's modulus
G	Shear modulus
I_y, I_z	Second moments of area about principal axes
I_0	Polar second moment of area about shear centre
J	St Venant torsion constant
M	Moment
M_{cr}	Critical moment including pre-buckling displacements
M_{cr0}	Critical moment: classical value
M_f	Moment in compression flange about major axis
M_{fp}	Plastic moment of compression flange about major axis
M_p	Full plastic moment of beam
M_u	Ultimate or failure moment of beam
P	Load
P_{cr}	Critical load including pre-buckling displacements
P_{cr0}	Critical load: classical value
P_f	Resultant axial force in compression flange
P_{fs}	Squash load of compression flange
P_i	Generalised external force
P_s	Squash load: area × yield stress
P_u	Ultimate or failure load
V	Total potential energy
α_w	Parameter relating to warping of cross-section
γ	Shear strain due to St Venant torsion
δ	First variation
δ^2	Second variation

ε	Axial strain
ε_i	Generalised strains
θ	Clockwise rotation about shear centre
θ_0	Initial value of θ
θ_{0c}	Initial value of θ at centre of beam
λ	Square root of M_p/M_{cr}
ξ	Distance from and normal to mean profile of cross-section
ρ	Distance from shear centre
σ	Axial stress
σ_i	Generalised stress
τ	Shear stress due to St Venant torsion
Δ	Small increment
[ ]	Square matrix
$\{\delta q\}$	Single column matrix of nodal displacement variations
$[KL]$	Linear stiffness matrix
$[KG]$	Geometric stiffness matrix: stability problems
$[KNL]$	Geometric stiffness matrix: non-linear problems

5.1 INTRODUCTION

Thin-walled bars of open cross-section such as angles, channels, T- and I-sections are used extensively in practice since they provide an economical use of material for the provision of stiffness and strength. For example, an I-section beam possesses much greater bending stiffness about its principal axes than a beam of solid square cross-section having the same total cross-section area. Such members can also be bolted or welded together to produce complete structures with relative ease.

The structural behaviour of thin-walled bars of open cross-section is, however, very complex since they are susceptible to instability in a variety of modes, and therefore to geometrically non-linear behaviour arising from initial imperfections. The various modes of instability of a thin-walled I-beam or beam column are illustrated in Fig. 5.1, each mode depending on the loading and support conditions. Figure 5.1(a) illustrates flexural instability about either the minor or the major principal axis and Fig. 5.1(b) illustrates pure torsional instability. Lateral instability, as illustrated in Fig. 5.1(c), is often referred to as flexural–torsional instability since it is a

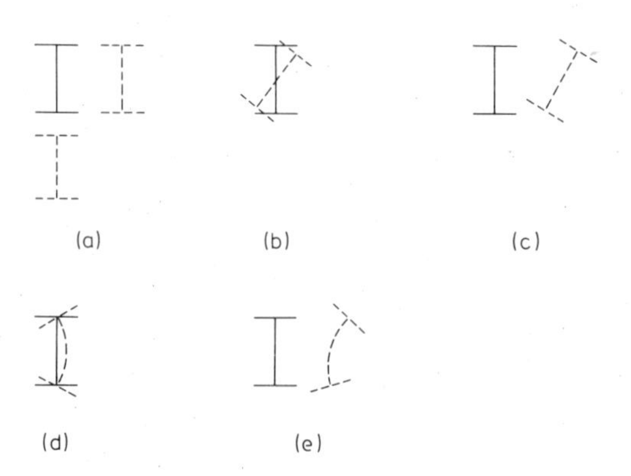

FIG. 5.1. Buckling modes of thin-walled I-beams.

combination of modes (a) and (b). Local instability, as illustrated in Fig. 5.1(d), results in distortion of the cross-section, without overall flexural and torsional instability. Distortional instability, as illustrated in Fig. 5.1(e), is a combination of local, flexural and torsional instability.

Elastic torsional and lateral instability has received much attention in the past and closed form solutions now exist for many problems involving relatively simple loading and boundary conditions (Timoshenko and Gere, 1961; Bleich, 1952; Chajes, 1974). The finite element method has been used to solve problems with more complex loading and support conditions. The finite element idealisations are based on either plate theory approximations (Nethercot and Rockey, 1971; Johnson and Will, 1974; Akay *et al.*, 1974) or beam theory approximations (Barsoum and Gallagher, 1970; Nethercot and Rockey, 1973; Nethercot, 1975). Solutions based on plate theory are able to take account of shear deformation and distortion of the cross-section. However, for many practical situations such considerations are not necessary and the analysis becomes relatively time-consuming and expensive when compared with solutions based on beam theory.

In the classical analysis of flexural–torsional instability it is assumed that the pre-buckling displacements are small enough to be neglected in the derivation of the governing differential equations. It has been shown, however, that this assumption is valid only when the ratios of the minor-axis flexural stiffness and torsional stiffness to the major-axis flexural stiffness are small (Davidson, 1952; Clark and Knoll, 1958; Vacharajittiphan *et al.*, 1974). When these ratios are not small, buckling loads obtained by inclusion of the pre-buckling displacements in the governing differential equations may exceed significantly the classical values for some practical rolled steel column sections. The analysis of the influence of pre-buckling displacements has been based mainly on the solution of the governing differential equations. A few closed form solutions have been obtained for simple loading and boundary conditions but, in general, numerical methods have to be used.

The geometrically non-linear behaviour of thin-walled bars of open cross-section has also been investigated. Masur and Milbradt (1957) have shown both theoretically and experimentally that the strength of redundant elastic narrow rectangular beams increases significantly after flexural–torsional buckling. The increase in strength occurs after only small lateral and torsional displacements, owing to axial straining which occurs since the ends of the beam are restrained. The post-buckled behaviour of statically determinate beams is quite different. Since the ends of the beam are not restrained, they pull in, no significant axial straining occurs, and such beams show no significant post-buckled increase in strength for small displacements. Woolcock and Trahair (1974) have shown, however, that very slender, simply supported and cantilever beams do show a post-buckled increase in strength at very large deformations. Their numerical solution of the governing non-linear differential equations was confirmed experimentally. The post-buckled increase in strength of statically determinate beams is not due to axial straining induced by end restraint but is a consequence of using the correct expressions for curvatures, instead of the usual linear approximations, and allowing for changes of geometry when deriving the governing differential equations. The behaviour is similar to that of the Elastica (Timoshenko and Gere, 1961). Typical load (P) vs. lateral displacement (v) curves obtained by Woolcock and Trahair (1974) for initially perfect and imperfect beams are shown in Fig. 5.2. In recent years the development of powerful numerical techniques such as the

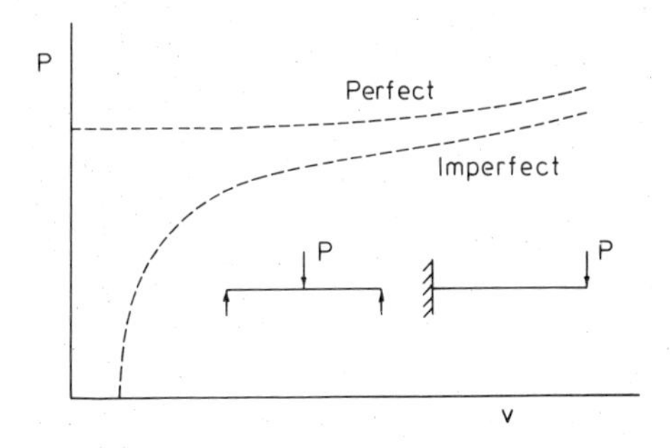

FIG. 5.2. Load vs. lateral deflection curves.

finite element method have facilitated a comprehensive theoretical study of the non-linear behaviour of thin-walled bars (Chen and Atsuta, 1977). In particular, Rajasekaran (1971, 1977) developed a general finite element algorithm for the geometrically non-linear, elasto-plastic analysis of such structures and many problems of practical interest can now be solved.

It is now almost universally accepted that elastic critical loads have little direct significance in predicting the failure loads of real structures. Simple structures such as beams and columns may fail at loads significantly below their elastic critical load while thin plated structures may be capable of sustaining loads several times greater than their elastic critical load. Elastic critical loads are used therefore mainly as reference loads, relative to which actual failure loads or general patterns of behaviour can be assessed. Since all real structures have small geometric imperfections, it is necessary to consider both geometric and material non-linearities in a realistic analysis.

The failure of struts and axially loaded columns has, for many years, been represented by Perry curves which relate the failure load to the slenderness ratio of the strut (Ayrton and Perry, 1886; Robertson, 1925). The slenderness ratio is defined as the effective length of the strut l_e, this being the distance between points of contraflexure, divided by the radius of gyration of the cross-section, r. A typical non-dimensional plot is shown in Fig. 5.3 in which P_u is the failure or ultimate load, P_s is the squash load (cross-section area $\times$ material yield stress) and P_{cr} is the elastic critical load for flexural buckling of a strut of effective length l_e. The position of the Perry curve depends upon the magnitude of the assumed initial geometric imperfection of the strut which can be adjusted to provide a mean or lower bound solution for experimental failure loads. In recent years the representation of failure loads has been refined, resulting in the new British and European column curves which depend not only on the assumed initial imperfection but also on the geometry of the cross-section (Dwight, 1978).

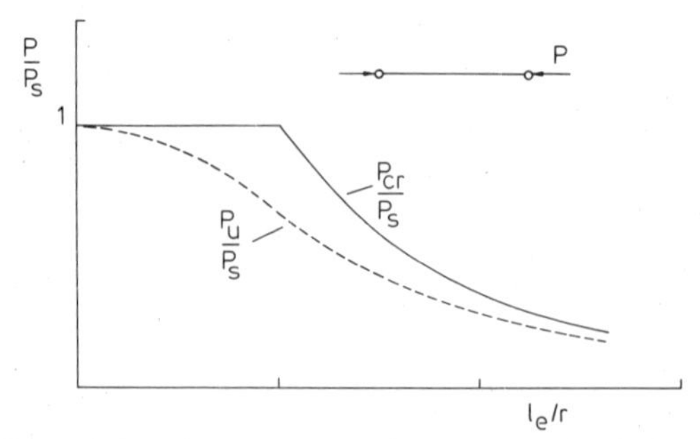

FIG. 5.3. Failure envelope for struts and columns.

The important feature of all these curves is that they predict failure at loads significantly below the elastic critical load, for a wide range of slenderness ratios.

The lateral failure of beams can be represented in a similar manner by plotting the failure moments M_u against a non-dimensional parameter $\lambda = (M_p/M_{cr})^{0 \cdot 5}$, where M_p is the plastic moment of resistance of the beam and M_{cr} is the elastic critical moment for flexural–torsional instability (Fukumoto and Kubo, 1977; Nethercot, 1978). The parameter λ was adopted since it contains all the material and geometric properties of the beam. A typical plot is shown in Fig. 5.4. Unlike the column curves, the curves representing the failure moment of beams have not been defined theoretically but are empirical curves which provide a mean or lower bound solution for experimental results. Lateral failure of beams occurs at moments significantly below the elastic critical moment for flexural–torsional buckling for a wide range of the parameter λ.

Non-linear expressions for the strains occurring in thin-walled bars of open cross-section when subjected to axial flexural and torsional displacements have been derived by Roberts (1981). The non-linear expressions for the strains were incorporated in a general instability analysis based on the vanishing of the second variation of the total potential energy. It was shown that after neglect of third- and higher-order terms in the displacement derivatives which appeared in the resulting energy equations, the classical solutions for flexural, torsional and lateral instability are reproduced. This work has been extended by Roberts and Azizian (1982*a*, *b*, 1983), using finite element techniques, to study the instability, including the influence of pre-buckling displacements, geometric non-linearity and collapse of such structures. These recent developments, and the results obtained, are discussed in this chapter.

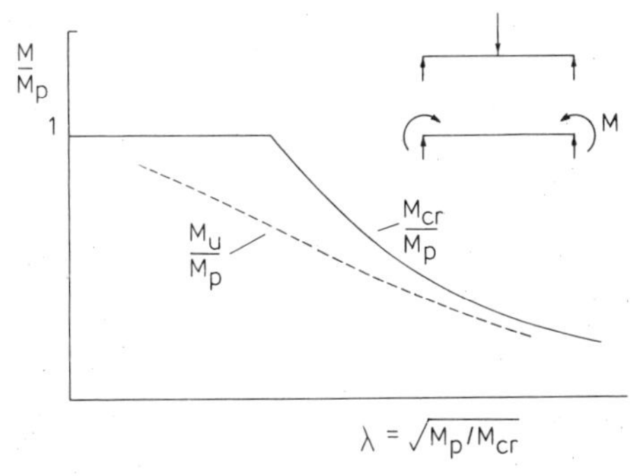

FIG. 5.4. Failure envelope for beams.

5.2 NON-LINEAR STRAINS

In the present theory, shear deformation due to non-uniform bending, and distortion of the cross-section, as illustrated in Fig. 5.1(d) and (e), are not considered.

A thin-walled bar of open cross-section is shown in Fig. 5.5(a). The x, y, z axes pass through the shear centre of the section which is denoted by O and are parallel to the principal axes which pass through the centroid C. The shear centre is the point through which the resultant shear force acts when the bar is subjected to non-uniform bending. The centre of rotation of the section is defined as the point about which the section rotates when subjected to a pure torque in the y–z plane and is coincident with the shear centre for linear strains. u denotes the displacement of the centroid in the x direction, v and w denote displacements of the shear centre in the y and z directions, and θ denotes the clockwise rotation of the section about the shear centre.

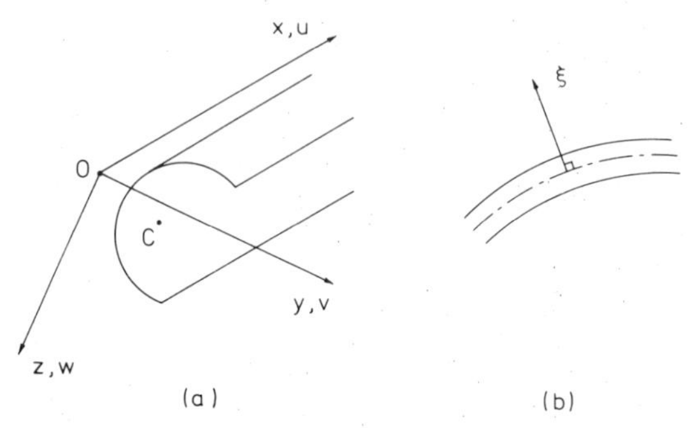

FIG. 5.5. Thin-walled bar with open cross-section.

The displacements u, v and w and the rotation θ produce non-linear axial and shear strains in the bar which vary over the cross-section. The derivation of these strains has been presented by Roberts (1981) and only the final expressions are reproduced here.

Letting suffixes x, xx and xxx denote differentiation with respect to x, the axial strain ε is given by

$$\begin{aligned}\varepsilon = & -y\{v_{xx}\,.\cos\theta + w_{xx}\,.\sin\theta\} - z\{w_{xx}\,.\cos\theta - v_{xx}\,.\sin\theta\} \\ & + \alpha_w\{\theta_{xx} - w_{xxx}\,.\,v_x + v_{xxx}\,.\,w_x\} \\ & + u_x + 0{\cdot}5\{v_x^2 + w_x^2 + \rho^2\theta_x^2\}\end{aligned} \tag{5.1}$$

The shear strain γ due to St Venant torsion is given by

$$\gamma = 2\xi\{\theta_x - w_{xx} \, . \, v_x + v_{xx} \, . \, w_x\} \tag{5.2}$$

In eqn (5.1) y and z are distances measured from the centroid in the positive y and z directions and ρ is the distance from the shear centre to any point on the cross-section. α_w is a parameter which relates to the warping of the section (Timoshenko and Gere, 1961; Roberts, 1981). In eqn (5.2) ξ is the distance measured from and perpendicular to the mean profile of the cross-section as shown in Fig. 5.5(b).

If the bar has small initial imperfections defined by v_0, w_0 and θ_0 and u, v, w and θ are total displacements (including initial imperfections), the non-linear expressions for ε and γ become

$$\begin{aligned}\varepsilon = & -y\{(v_{xx} \, . \cos\theta - v_{0xx} \, . \cos\theta_0) + (w_{xx} \, . \sin\theta - w_{0xx} \, . \sin\theta_0)\} \\ & - z\{(w_{xx} \, . \cos\theta - w_{0xx} \, . \cos\theta_0) - (v_{xx} \, . \sin\theta - v_{0xx} \, . \sin\theta_0)\} \\ & + \alpha_w\{(\theta_x - \theta_{0x}) - (w_{xxx} \, . \, v_x - w_{0xxx} \, . \, v_{0x}) + (v_{xxx} \, . \, w_x - v_{0xxx} \, . \, w_{0x})\} \\ & + u_x + 0{\cdot}5\{(v_x^2 - v_{0x}^2) + (w_x^2 - w_{0x}^2) + \rho^2(\theta_x^2 - \theta_{0x}^2)\}\end{aligned} \tag{5.3}$$

$$\gamma = 2\xi\{(\theta_x - \theta_{0x}) - (w_{xx} \, . \, v_x - w_{0xx} \, . \, v_{0x}) + (v_{xx} \, . \, w_x - v_{0xx} \, . \, w_{0x})\} \tag{5.4}$$

5.3 ENERGY EQUATIONS

5.3.1 Instability

Equilibrium states are defined by the condition that the first variation of the total potential energy V, denoted by δV, is zero (Langhaar, 1962; Roberts, 1981). For a structure subjected to external forces P_i, the corresponding displacements being r_i, this statement can be represented by the equation (external forces being assumed stationary in the equilibrium state)

$$\delta V = -P_i \, . \, \delta r_i + \int \sigma_i \, . \, \delta\varepsilon_i \, . \, \mathrm{d\,vol} = 0 \tag{5.5}$$

in which repeated suffixes imply summation and the integration is over the volume of the structure. σ_i represents the stresses in the structure in the equilibrium state and $\delta\varepsilon_i$ represents the first variation in the corresponding strains.

For stable equilibrium, $\delta V = 0$ corresponds to a minimum value and the second variation of V, denoted by $\delta^2 V$, is positive definite, i.e. positive for all admissible variations in displacements and corresponding strains. In a

Taylor's series expansion of a function F of several displacement variables q_i, first and second variations are defined as

$$\delta F = \frac{\partial F}{\partial q_i} \cdot \delta q_i \qquad \delta^2 F = \frac{1}{2} \frac{\partial^2 F}{\partial q_i \partial q_j} \cdot \delta q_i \delta q_j \tag{5.6}$$

Hence from eqn (5.5), for stationary values of the applied loads, $\delta^2 V$ is given by

$$\delta^2 V = -P_i \delta^2 r_i + \int \{\sigma_i \delta^2 \varepsilon_i + 0{\cdot}5 \delta\sigma_i \delta\varepsilon_i\} \, \mathrm{d\,vol} \tag{5.7}$$

Critical conditions occur when $\delta^2 V$ vanishes, indicating a possible transition from stability to instability. Since the second variation of any linear function vanishes, it is necessary to consider non-linear strains to define (5.7).

5.3.2 Geometric Non-Linearity: Incremental Analysis

For a structure in equilibrium, the first variation of the total potential energy, defined by eqn (5.5), is zero. Equation (5.5) can be rewritten as

$$P_i \delta r_i = \int \sigma_i \delta\varepsilon_i \, \mathrm{d\,vol} \tag{5.8}$$

If Δ denotes small but finite increments in forces, displacements, stresses and strains such that the structure moves to an adjacent equilibrium state, then

$$(P_i + \Delta P_i)\delta(r_i + \Delta r_i) = \int (\sigma_i + \Delta\sigma_i)\delta(\varepsilon_i + \Delta\varepsilon_i) \, \mathrm{d\,vol} \tag{5.9}$$

In eqn (5.9) r_i and ε_i can be considered as constants which define the previous equilibrium state and hence eqn (5.9) can be rewritten as

$$(P_i + \Delta P_i)\delta\Delta r_i = \int (\sigma_i + \Delta\sigma_i)\delta\Delta\varepsilon_i \, \mathrm{d\,vol} \tag{5.10}$$

Subtracting eqn (5.8) from eqn (5.10) gives

$$P_i(\delta\Delta r_i - \delta r_i) + \Delta P_i \delta\Delta r_i = \int \sigma_i(\delta\Delta\varepsilon_i - \delta\varepsilon_i) \, \mathrm{d\,vol} + \int \Delta\sigma_i \delta\Delta\varepsilon_i \, \mathrm{d\,vol} \tag{5.11}$$

Equation (5.11) provides a basis for incremental analysis of non-linear problems.

For many problems the displacements r_i of the external forces can be expressed as linear functions of displacement variables. It may then be assumed that $\delta r_i \equiv \delta\Delta r_i$ since eqns (5.8) and (5.10) are valid for all admissible variations in displacements and corresponding strains. The first term in eqn (5.11) can then be omitted.

For the particular problem under consideration, the strains ε_i (axial strain ε and shear strain γ) are functions of displacements u, v, w and θ. Since we are considering non-linear strains, the strain increments $\Delta\varepsilon_i$ are functions of displacements u, v, w and θ and displacement increments Δu, Δv, Δw and $\Delta\theta$. If displacements and displacement increments can be expressed as linear functions of displacement variables it can be assumed, for the same reason as stated in the previous paragraph, that

$$\delta u \equiv \delta\Delta u \qquad \delta v \equiv \delta\Delta v \qquad \delta w \equiv \delta\Delta w \qquad \delta\theta \equiv \delta\Delta\theta \tag{5.12}$$

When considering non-linear strains, it cannot be assumed that $\delta\varepsilon_i \equiv \delta\Delta\varepsilon_i$.

5.4 INSTABILITY ANALYSIS

The non-linear strains defined in Section 5.2 and the energy equations derived in Section 5.3.1 facilitate a study of the instability of thin-walled bars.

In eqn (5.7) σ_i represents the axial stress σ and shear stress τ in the bar prior to buckling. These stresses can usually be expressed as linear functions of the pre-buckling displacements and for elastic material the most general forms are

$$\sigma = E\{u_x - zw_{xx} - yv_{xx} + \alpha_w\theta_{xx}\} \tag{5.13}$$

$$\tau = G2\xi\theta_x \tag{5.14}$$

in which E is Young's modulus and G is the shear modulus. Equation (5.7) can now be rewritten as

$$\delta^2 V = -P_i\delta^2 r_i + \int\{\sigma\,.\,\delta^2\varepsilon + \tau\,.\,\delta^2\gamma + 0{\cdot}5(\delta\sigma\,.\,\delta\varepsilon + \delta\tau\,.\,\delta\gamma)\}\,\mathrm{d\,vol} \tag{5.15}$$

In eqn (5.15) $\delta\sigma$ and $\delta\tau$ are the first variations of the axial and shear strains which are related to $\delta\varepsilon$ and $\delta\gamma$ by the equations

$$\delta\sigma = E\,\delta\varepsilon \qquad \delta\tau = G\,\delta\gamma \tag{5.16}$$

Hence eqn (5.15) becomes

$$\delta^2 V = -P_i\delta^2 r_i + \int\{\sigma\,.\,\delta^2\varepsilon + \tau\,.\,\delta^2\gamma + 0{\cdot}5(E\delta\varepsilon^2 + G\delta\gamma^2)\}\,\mathrm{d\,vol} \tag{5.17}$$

The first and second variations of the axial and shear strains, given by eqns (5.1) and (5.2), are

$$\delta\varepsilon = -y\{-v_{xx}.\sin\theta.\delta\theta+\delta v_{xx}.\cos\theta+w_{xx}.\cos\theta.\delta\theta+\delta w_{xx}.\sin\theta\}$$
$$-z\{-w_{xx}.\sin\theta.\delta\theta+\delta w_{xx}.\cos\theta-v_{xx}.\cos\theta.\delta\theta-\delta v_{xx}.\sin\theta\}$$
$$+\alpha_{w}\{\delta\theta_{xx}-w_{xxx}.\delta v_{x}-\delta w_{xxx}.v_{x}+v_{xxx}.\delta w_{x}+\delta v_{xxx}.w_{x}\}$$
$$+\delta u_{x}+v_{x}.\delta v_{x}+w_{x}.\delta w_{x}+\rho^{2}\theta_{x}.\delta\theta_{x} \qquad (5.18)$$

$$\delta^{2}\varepsilon = -\frac{y}{2}\{-v_{xx}.\cos\theta.\delta\theta.\delta\theta-2\delta v_{xx}.\sin\theta.\delta\theta-w_{xx}.\sin\theta.\delta\theta.\delta\theta$$
$$+2\delta w_{xx}.\cos\theta.\delta\theta\}-\frac{z}{2}\{-w_{xx}.\cos\theta.\delta\theta.\delta\theta-2\delta w_{xx}.\sin\theta.\delta\theta$$
$$+v_{xx}.\sin\theta.\delta\theta.\delta\theta-2\delta v_{xx}.\cos\theta.\delta\theta\}$$
$$+\frac{\alpha_{w}}{2}\{-2\delta w_{xxx}.\delta v_{x}+2\delta v_{xxx}.\delta w_{x}\}$$
$$+0{\cdot}5\{\delta v_{x}.\delta v_{x}+\delta w_{x}.\delta w_{x}+\rho^{2}\delta\theta_{x}.\delta\theta_{x}\} \qquad (5.19)$$

$$\delta\gamma = 2\xi\{\delta\theta_{x}-w_{xx}.\delta v_{x}-\delta w_{xx}.v_{x}+v_{xx}.\delta w_{xx}+\delta v_{xx}.w_{x}\} \qquad (5.20)$$

$$\delta^{2}\gamma = 2\xi\{-\delta w_{xx}.\delta v_{x}+\delta v_{xx}.\delta w_{x}\} \qquad (5.21)$$

Terms such as $\delta^2 u$, $\delta^2 v$, etc., do not appear in the second variation since it is assumed that u, v, etc., can be expressed as linear functions of displacement variables and hence that $\delta^2 u$, $\delta^2 v$, etc., vanish.

Substituting eqns (5.13), (5.14) and (5.18)–(5.21) into eqn (5.17) gives the complete expression for $\delta^2 V$. Critical conditions occur when $\delta^2 V = 0$ and solutions can be obtained by arranging $\delta^2 V$ as a complete quadratic form and equating the determinant of the quadratic form to zero (Roberts, 1981).

To simplify eqn (5.17), the integration over the volume is replaced by the integral $\mathrm{d}A.\mathrm{d}x$, in which A is the cross-section area of the bar. The following integrals are then used:

$$\int y\,\mathrm{d}A = \int z\,\mathrm{d}A = \int yz\,\mathrm{d}A = 0 \qquad (5.22)$$

$$\int \alpha_{w}\,\mathrm{d}A = \int y\alpha_{w}\,\mathrm{d}A = \int z\alpha_{w}\,\mathrm{d}A = 0 \qquad (5.23)$$

$$\int \mathrm{d}A = A \qquad \int y^{2}\,\mathrm{d}A = I_{z} \qquad \int z^{2}\,\mathrm{d}A = I_{y} \qquad \int \rho^{2}\,\mathrm{d}A = I_{0} \qquad (5.24)$$

$$\int \alpha_{w}^{2}\,\mathrm{d}A = C_{w} \qquad \int 4\xi^{2}\,\mathrm{d}A = J \qquad (5.25)$$

$$\int y\rho^{2}\,\mathrm{d}A = C_{z} \qquad \int z\rho^{2}\,\mathrm{d}A = C_{y} \qquad \int \alpha_{w}\rho^{2}\,\mathrm{d}A = C_{\rho w} \qquad \int \rho^{4}\,\mathrm{d}A = C_{\rho} \qquad (5.26)$$

Equations (5.22) are the first moments of area and product second moment of area about the centroid. Equations (5.23) are a consequence of the reciprocal theorem. A is the cross-section area, I_z and I_y are the second moments of area about the principal axes and I_0 is the polar second moment of area about the shear centre. C_w is the warping constant and J is the St Venant torsion constant. C_y, C_z, $C_{\rho w}$ and C_ρ represent the influence of second-order axial strains induced by rotation about the shear centre. C_y and C_z are zero if the y and z axes respectively are axes of symmetry. $C_{\rho w}$ and C_ρ are associated only with pre-buckling rotations and hence do not appear in the problems of instability discussed herein.

5.4.1 Symmetrical I-Beam under Axial Compression

The problem under consideration is illustrated in Fig. 5.6. For a symmetrical I-beam the shear centre coincides with the centroid. The length of the beam is a. The boundary conditions are $u = 0$ at $x = 0$, and at $x = 0$ and a,

$$v = w = \theta = v_{xx} = w_{xx} = \theta_{xx} = 0 \tag{5.27}$$

The potential of the applied loads P, which correspond to surface tractions $T = -P/A$ (tensile positive) uniformly distributed over the ends of the beam, is

$$-[\textstyle\int -Tu\,\mathrm{d}A]_{x=0} - [\textstyle\int Tu\,\mathrm{d}A]_{x=a} \tag{5.28}$$

Since this is a linear function of the displacements u occurring at $x = 0$ and a, the second variation vanishes and hence the term $-P_i\delta^2 r_i$ in eqn (5.17) vanishes also. The stress in the beam prior to buckling, given in general by eqns (5.13) and (5.14), can be replaced simply by $\sigma = -P/A$.

The complete expression for $\delta^2 V$, which vanishes for critical conditions, then becomes

$$\delta^2 v = \int_0^a -0{\cdot}5P\left\{(\delta v_x)^2 + (\delta w_x)^2 + \frac{I_0}{A}(\delta\theta_x)^2\right\}\mathrm{d}x$$
$$+0{\cdot}5\int_0^a \{EI_z(\delta v_{xx})^2 + EI_y(\delta w_{xx})^2 + EC_w(\delta\theta_{xx})^2$$
$$+GJ(\delta\theta_x)^2 + EA(\delta u_x)^2\}\,\mathrm{d}x = 0 \tag{5.29}$$

Equation (5.29) can be solved by assuming suitable displacement functions for δv, δw and $\delta\theta$, δu being assumed zero since the term containing δu_x is positive definite and can therefore only increase the calculated critical value of P.

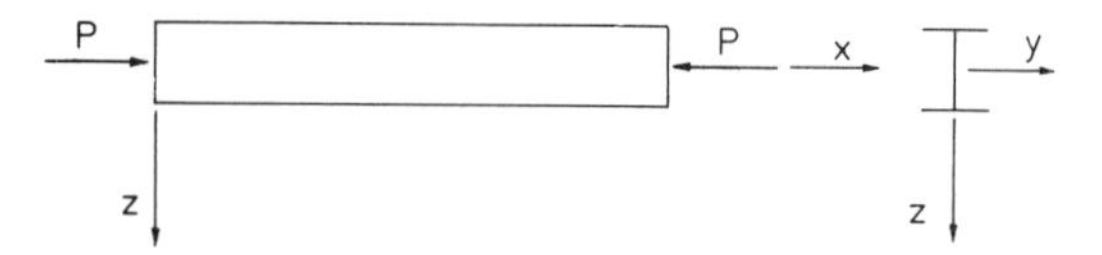

FIG. 5.6. Axially loaded beam.

Displacement functions which satisfy the prescribed boundary conditions are:

$$\begin{aligned} \delta v &= C_1 \sin \pi x/a \\ \delta w &= C_2 \sin \pi x/a \\ \delta\theta &= C_3 \sin \pi x/a \end{aligned} \tag{5.30}$$

where C_1, C_2 and C_3 define the absolute magnitude of the displacement variations. Substituting these expressions into eqn (5.29) gives

$$-\frac{P}{2}\frac{\pi^2}{a^2}\left\{C_1^2 + C_2^2 + C_3^2\frac{I_0}{A}\right\} + \frac{1}{2}\frac{\pi^4}{a^4}\left\{EI_z C_1^2 + EI_y C_2^2 + \left(EC_w + GJ\frac{a^2}{\pi^2}\right)C_3^2\right\} = 0 \tag{5.31}$$

Equation (5.31) can be arranged in matrix form as

$$0{\cdot}5[C_1 \;\; C_2 \;\; C_3]\begin{bmatrix} \dfrac{\pi^2 EI_z}{a^2} - P & 0 & 0 \\ 0 & \dfrac{\pi^2 EI_y}{a^2} - P & 0 \\ 0 & 0 & \dfrac{\pi^2 EC_w}{a^2} + GJ - \dfrac{PI_0}{A} \end{bmatrix}\begin{bmatrix} C_1 \\ C_2 \\ C_3 \end{bmatrix} = 0 \tag{5.32}$$

Equation (5.32) is a complete quadratic form which changes from positive definite to semi-positive definite when the determinant vanishes. Therefore, critical conditions occur when

$$\left(\frac{\pi^2 EI_z}{a^2} - P\right)\left(\frac{\pi^2 EI_y}{a^2} - P\right)\left(\frac{\pi^2 EC_w}{a^2} + GJ - \frac{PI_0}{A}\right) = 0 \tag{5.33}$$

The three roots of eqn (5.33) are the critical loads which correspond to pure flexural buckling about the y and z axes and to pure torsional buckling (Timoshenko and Gere, 1961). Combined flexural and torsional buckling does not occur in this particular case.

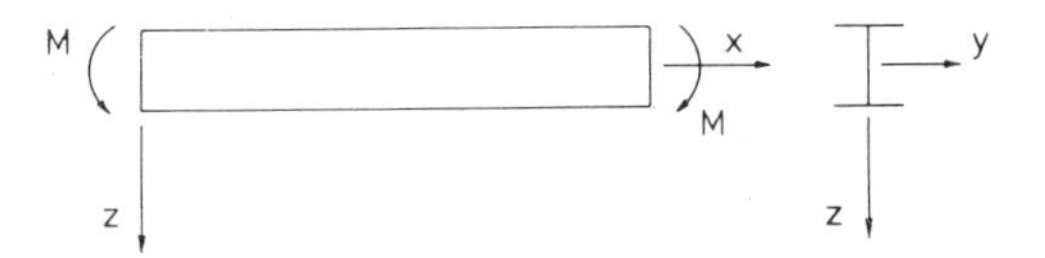

FIG. 5.7. Beam subjected to uniform bending.

5.4.2 Symmetrical I-Beam Subjected to Uniform Bending

The problem is illustrated in Fig. 5.7. The potential of the applied moments is

$$-[-Mw_x]_{x=0} - [Mw_x]_{x=a} \tag{5.34}$$

and since w_x is linear, the second variation vanishes. The moments produce stresses in the beam prior to buckling which vary linearly throughout the depth and which are given by

$$\sigma = \frac{Mz}{I_y} \tag{5.35}$$

The boundary conditions are $u = 0$ at $x = 0$, and at $x = 0$ and a,

$$v = w = \theta = v_{xx} = w_{xx} = \theta_{xx} = 0 \tag{5.36}$$

If the pre-buckling displacements are assumed small and therefore neglected, $\delta^2 V = 0$ reduces to

$$\int_0^a -M\,\delta v_{xx}\,.\,\delta\theta\,\mathrm{d}x + 0{\cdot}5 \int_0^a \{EI_z(\delta v_{xx})^2 + EI_y(\delta w_{xx})^2 + EC_{\mathrm{w}}(\delta\theta_{xx})^2 + GJ(\delta\theta_x)^2 + EA(\delta u_x)^2\}\,\mathrm{d}x = 0 \tag{5.37}$$

Assuming displacement variations δu and δw to be zero, since terms containing u and w in eqn (5.37) are positive definite, a solution can be obtained by assuming

$$\begin{aligned} \delta v &= C_1 \sin \pi x/a \\ \delta\theta &= C_2 \sin \pi x/a \end{aligned} \tag{5.38}$$

Substituting these expressions into eqn (5.37) and proceeding as in the previous section, the critical moment for lateral or flexural–torsional buckling, neglecting the influence of pre-buckling displacements, is given by Timoshenko and Gere (1961):

$$M_{\mathrm{cr0}} = \frac{\pi}{a}\sqrt{\left(EI_z GJ\left\{1 + \frac{EC_{\mathrm{w}}}{GJ}\left(\frac{\pi}{a}\right)^2\right\}\right)} \tag{5.39}$$

5.4.3 Symmetrical I-Beam with a Concentrated Load Acting at the Centre

The problem is illustrated in Fig. 5.8 and the boundary conditions are as in Section 5.4.2. If the pre-buckling displacements are neglected, $\delta^2 V = 0$ reduces to

$$0{\cdot}5 \int_0^{a/2} P\,.\,x\,.\,\delta v_{xx}\,.\,\delta\theta\,.\,\mathrm{d}x + 0{\cdot}5 \int_0^{a/2} \{EI_z(\delta v_{xx})^2 + EI_y(\delta w_{xx})^2 + EC_w(\delta\theta_{xx})^2 + GJ(\delta\theta_x)^2 + EA(\delta u_x)^2\}\,\mathrm{d}x = 0 \tag{5.40}$$

Assuming displacement variations as in Section 5.4.2, the critical load for lateral or flexural–torsional buckling, neglecting the influence of pre-buckling displacements, is given by Chajes (1974):

$$P_{\mathrm{cr0}} = 4\left(\frac{\pi}{a}\right)^2 \sqrt{\left(\frac{3}{\pi^2+6}\, EI_z \left\{GJ + EC_w \left(\frac{\pi}{a}\right)^2\right\}\right)} \tag{5.41}$$

If the load P is applied on the top flange, the displacement of the load contains second-order terms due to the rotation and hence the second variation of the potential of the applied loads does not vanish. The theory then has to be modified slightly as discussed by Roberts (1981) and Roberts and Azizian (1983).

5.4.4 Influence of Pre-Buckling Displacements

If the pre-buckling displacements are not neglected in the expression for the second variation of the total potential energy, the elastic critical loads so determined may be significantly greater than the well-known classical values.

For the problem of a symmetrical I-beam subjected to uniform bending, as discussed in Section 5.4.2, the complete expression for $\delta^2 V = 0$ becomes (Roberts and Azizian, (1982*a*):

$$\begin{aligned}
&\int \frac{M}{2}\{-2\delta v_{xx}\,.\,\delta\theta - w_{xx}\,.\,\delta\theta\,.\,\delta\theta\}\,\mathrm{d}x \\
&\quad + \frac{GJ}{2}\int \{\delta\theta_x - w_{xx}\,.\,\delta v_x + w_x\,.\,\delta v_{xx}\}^2 \\
&\quad + \frac{E}{2}\int \{I_z(\delta v_{xx} + w_{xx}\,.\,\delta\theta)^2 + I_y \delta w_{xx}\,.\,\delta w_{xx} \\
&\quad + C_w(\delta\theta_{xx} - w_{xxx}\,.\,\delta v_x + w_x\,.\,\delta v_{xxx})^2 \\
&\quad + A(\delta u_x\,.\,\delta u_x + w_x\,.\,w_x\,.\,\delta w_x\,.\,\delta w_x)\}\,\mathrm{d}x = 0
\end{aligned} \tag{5.42}$$

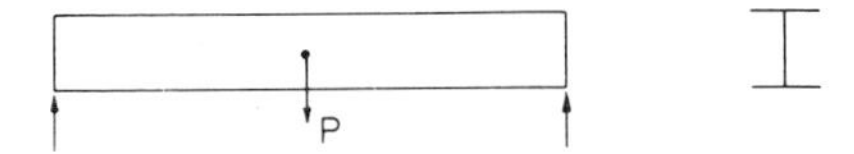

FIG. 5.8. Beam with central point load.

The pre-buckling displacements w can be expressed in terms of the applied moments M and a solution obtained as in Section 5.4.3. The critical moment for lateral buckling, including the influence of pre-buckling displacements, M_{cr}, is then given by

$$M_{cr} = \frac{\pi}{a}\sqrt{\left(\frac{EI_z GJ\left\{1 + \frac{EC_w}{GJ}\left(\frac{\pi}{a}\right)^2\right\}}{\left\{1 - \frac{I_z}{I_y}\right\}\left\{1 - \frac{GJ}{EI_y}\left[1 + \frac{EC_w}{GJ}\left(\frac{\pi}{a}\right)^2\right]\right\}}\right)} \tag{5.43}$$

Equation (5.43) agrees exactly with the solution based on the governing differential equations obtained by Vacharajittiphan *et al.* (1974).

The dominant term in the denominator of eqn (5.43) is $\{1 - I_z/I_y\}$ and the percentage increase in the critical moment due to pre-buckling displacements is almost independent of the span. If the web contribution to I_y and I_z is neglected, the ratio I_z/I_y is approximately proportional to $(b/d)^2$, where b is the flange width and d is the depth of the beam. It is convenient therefore to plot the increase ratio M_{cr}/M_{cr0} against b/d and such a plot is shown in Fig. 5.9.

For the problem of a symmetrical I-beam with a concentrated load acting at the centre, as discussed in Section 5.4.3, the obtaining of closed form solutions is possible but the analysis and resulting expressions are

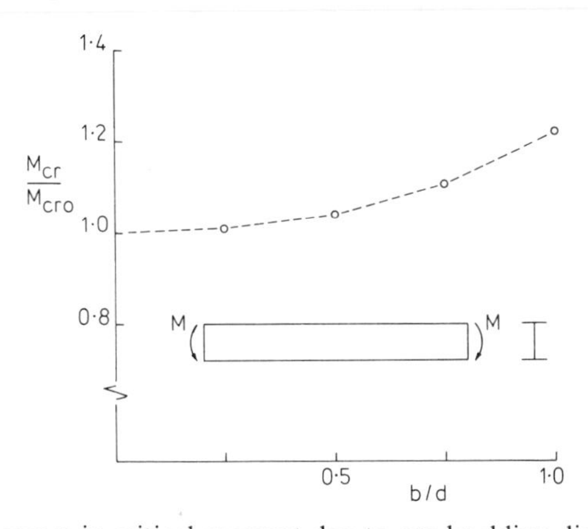

FIG. 5.9. Increase in critical moment due to pre-buckling displacements.

lengthy. Solutions have been obtained therefore by Roberts and Azizian (1983) using well-known finite element techniques (Gallagher and Padlog, 1963; Kapur and Hartz, 1966; Zienkiewicz, 1971). Using finite element techniques, $\delta^2 V = 0$ can be expressed as

$$\tfrac{1}{2}\{\delta q\}^T([KL] + [KG])\{\delta q\} = 0 \tag{5.44}$$

in which $\{\delta q\}$ is the vector of nodal displacement variations, $[KL]$ is the linear stiffness matrix for axial, flexural and torsional displacements and $[KG]$ is the geometric stiffness matrix which depends on the strains prior to buckling. Equation (5.44) is a complete quadratic form which changes from positive definite to semi-positive definite when the determinant of $[KL] + [KG]$ vanishes.

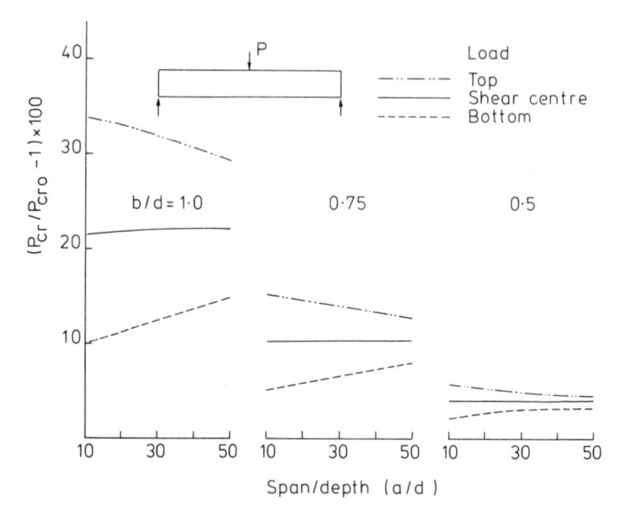

FIG. 5.10. Increase in critical load due to pre-buckling displacements.

The numerical analysis confirmed that the percentage increase in the ratio P_{cr}/P_{cr0}, where P_{cr} is the critical load given by the present theory and P_{cr0} is the classical value of the critical load given by eqn (5.41) and Timoshenko and Gere (1961), depends on the span/depth ratio a/d and also on the point of application of the load relative to the centre of rotation. The percentage increase in the ratio P_{cr}/P_{cr0} for different values of a/d and b/d is shown in Fig. 5.10.

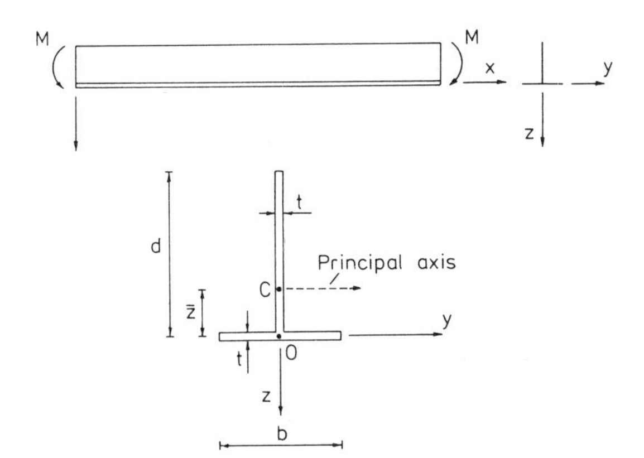

FIG. 5.11. Inverted T-beam subjected to uniform bending.

5.4.5 Unsymmetrical Sections

The application of the preceding theory for unsymmetrical sections is now illustrated by considering the instability of a simply supported inverted T-beam subjected to uniform bending, as shown in Fig. 5.11 (Roberts and Azizian, 1982*a*). The shear centre and centroid of the section are located at O and C respectively and α_w for the section is zero. The boundary conditions are as defined in Section 5.4.2 and the stresses in the beam prior to buckling are given by

$$\sigma = -\frac{Mz}{I_y} \tag{5.45}$$

The complete expression for $\delta^2 V = 0$, including the pre-buckling displacements w which can be expressed in terms of the applied moment, then becomes

$$\begin{aligned}\delta^2 V = &-\int \frac{M}{2}\left\{\frac{M}{EI_y}\cdot \delta\theta\,.\,\delta\theta + 2\delta v_{xx}\cdot \delta\theta + \frac{C_y}{I_y}\,\delta\theta_x\,.\,\delta\theta_x\right\}\mathrm{d}x \\ &+ \frac{GJ}{2}\int\left\{\delta\theta_x - \frac{2M}{EI_y}\cdot \delta\theta_x\,.\,\delta v_x + \frac{2M}{EI_y}(x-a/2)\delta\theta_x\,.\,\delta v_{xx}\right\}\mathrm{d}x \\ &+ \frac{EI_z}{2}\int\left\{\delta v_{xx}\cdot \delta v_{xx} + \frac{2M}{EI_y}\cdot \delta v_{xx}\,.\,\delta\theta + \left(\frac{M}{EI_y}\right)^2 \delta\theta\,.\,\delta\theta\right\}\mathrm{d}x = 0\end{aligned} \tag{5.46}$$

In arriving at eqn (5.46), positive definite terms have been omitted. C_y is defined in eqns (5.26) and is not zero since the y-axis is not an axis of symmetry.

Assuming the displacement variations defined in eqns (5.28) and that $GJ/EI_y \ll 1$, the solution for the critical moment can be expressed as

$$M_{cr} = \frac{-D \pm \sqrt{(D^2 - 4BH)}}{2B} \tag{5.47}$$

in which

$$D = \frac{EC_y I_z}{I_y}\left(\frac{\pi}{a}\right)^2$$
$$B = \left\{1 - \frac{I_z}{I_y}\right\} \tag{5.48}$$
$$H = -EI_z GJ\left(\frac{\pi}{a}\right)^2$$

Equation (5.47) predicts different critical values for positive and negative moments which is in accordance with existing solutions for unsymmetrical sections. If B, which represents the influence of pre-buckling displacements, is assumed equal to unity, eqn (5.47) agrees exactly with existing solutions for monosymmetric I-beams, when the width of one flange is reduced to zero (Anderson and Trahair, 1972).

5.5 GEOMETRICALLY NON-LINEAR ANALYSIS

The basic finite element algorithms for the geometrically non-linear analysis of structures were developed during the late sixties and early seventies and since that time many problems of practical interest have been solved (Schmit *et al.*, 1968; Murray and Wilson, 1968; Kawai and Yoshimura, 1969; Brebbia and Connor, 1969; Roberts and Ashwell, 1971; Zienkiewicz, 1971). Solutions can be obtained iteratively, incrementally or by a combination of increments and iterations.

Solutions of the present problem have been obtained incrementally by Roberts and Azizian (1982*b*). The basic energy equations for incremental analysis are given in Section 5.3.2. From eqn (5.11), and the non-linear strains defined in Section 5.2, it is possible to derive a set of simultaneous equations of the form

$$\{\Delta P\} = ([KL] + [KNL])\{\Delta q\} \tag{5.49}$$

relating increments in the nodal forces $\{\Delta P\}$ to increments in the nodal displacements $\{\Delta q\}$. $[KL]$ is the linear stiffness matrix for axial, flexural and torsional displacements and $[KNL]$ is the geometric stiffness matrix, similar to $[KG]$ in eqn (5.44), which depends on the current strain state. Solutions can be obtained by incrementing loads proportionally or by rearranging eqn (5.49) and incrementing one of the displacements. This latter approach is convenient when the load–deflection curves are expected to approach or pass through a horizontal tangent (Roberts and Ashwell, 1971).

5.5.1 Axially Loaded Column

This problem has been chosen to illustrate the accuracy of the finite element solution since exact analytical solutions are available (Chajes, 1974).

The problem under consideration is illustrated in Fig. 5.12. The column has an initial imperfection defined by

$$v_0 = v_{0c} \sin \frac{\pi x}{a} \tag{5.50}$$

The displacement boundary conditions are $w = \theta = 0$ everywhere, $u = v = 0$ at $x = 0$ and $v = 0$ at $x = a$, i.e. the axial force produces bending of the column about its minor axis. Sixteen elements were used to model the column and the results shown in Fig. 5.12 were obtained by incrementing v_c at the centre of the column. $P_{cr} = \pi^2 EI_z/a^2$ is the elastic critical load of the column for flexural buckling about the z-axis.

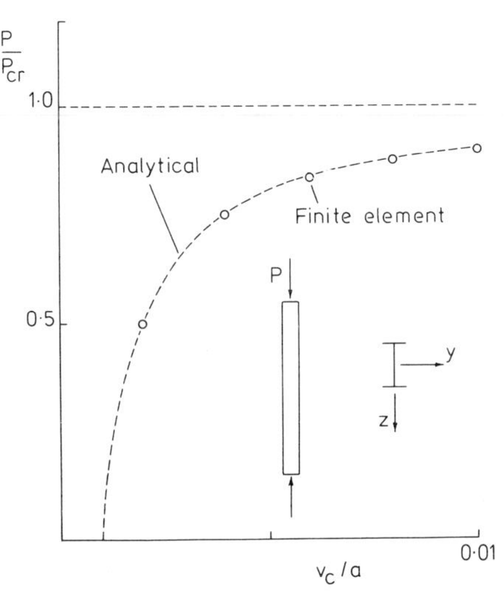

FIG. 5.12. Load deflection curve for pin-ended column.

5.5.2 I-Beam Subjected to Uniform Moment

The problem is illustrated in Fig. 5.13. The beam has initial imperfections defined by

$$v_0 = v_{0c} \sin \frac{\pi x}{a} \qquad \theta_0 = \theta_{0c} \sin \frac{\pi x}{a} \tag{5.51}$$

The displacement boundary conditions are $u = v = w = \theta = 0$ at $x = 0$ and $v = w = \theta = 0$ at $x = a$. Sixteen elements were used to model the beam and results were obtained by incrementing v at the centre of the beam.

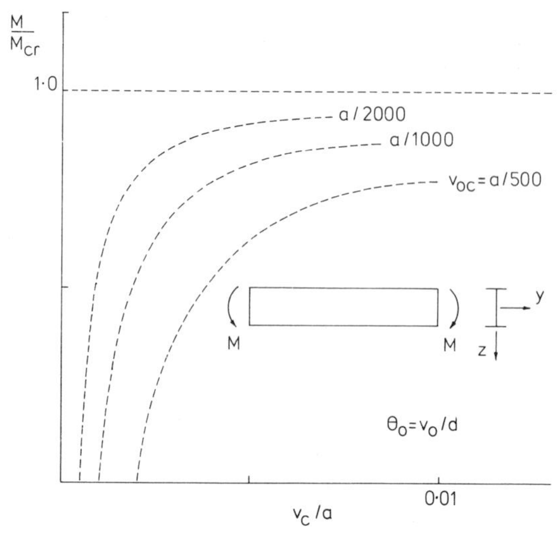

FIG. 5.13. Moment vs. lateral deflection curves for imperfect beams.

The moment vs. lateral deflection (v_c) curves for three different magnitudes of the initial imperfections are shown in Fig. 5.13. M_{cr} is the elastic critical moment for lateral buckling, including the influence of pre-buckling displacements, given by eqn (5.43). Similar curves to those shown in Fig. 5.13 are obtained for simply supported beams subjected to concentrated loads applied at the centre of the beam.

5.6 COLLAPSE ANALYSIS

The results obtained from a geometrically non-linear analysis can be used to predict the lateral failure of initially imperfect beams (Roberts and Azizian, 1982*b*). It is assumed that lateral failure occurs when the compression flange cannot sustain any increase in stress. The elastic stress distribution in the compression flange due to displacements u, v, w and θ can be integrated over the area of the flange to give a resultant axial force P_f and a resultant moment M_f about the major axis of the flange. The shear stresses in the flange due to St Venant torsion have been neglected in the present analysis. Failure is then assumed to occur when M_f reaches the reduced plastic moment of the flange about the major axis. For a rectangular flange, this can be represented by the equation

$$M_f = M_{fp}\left\{1 - \left(\frac{P_f}{P_{fs}}\right)^2\right\} \tag{5.52}$$

in which M_{fp} is the full plastic moment of the flange and P_{fs} is the squash load of the flange which is equal to the material yield stress × the cross-section area.

The results of such an analysis are shown in Fig. 5.14 for simply supported beams subjected to uniform bending and to a concentrated load applied on the top flange. In Fig. 5.14, M_u is the ultimate moment that the beam can sustain according to the assumed failure criterion, M_p is the plastic moment of resistance of the beam for bending about its major axis, and M_{cr} is the elastic critical moment (maximum moment in the beam corresponding to P_{cr} for a central concentrated load). Three curves are presented for three different magnitudes of the assumed initial imperfections. The failure envelopes so plotted appear independent of the loading

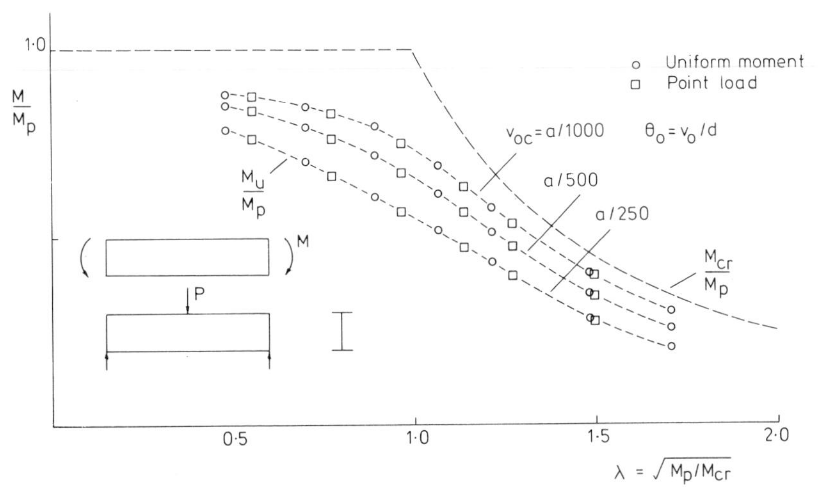

FIG. 5.14. Failure envelopes for beams.

condition and a more extensive study has shown them to be almost independent of the cross-section dimensions of the beam and of the material yield stress. This substantiates the use of the non-dimensional parameter $\lambda = (M_p/M_{cr})^{0.5}$ in representing failure envelopes.

The theoretical curves do not tend to $M_u/M_p = 1$ for $\lambda = 0$ owing to the assumed failure criterion. According to the assumed criterion, lateral failure occurs when the compression flange yields, which is a reasonable assumption since the beam then has little or no resistance to loading which induces lateral and torsional displacements.

In Fig. 5.15 the failure envelope for $v_{0c} = a/500$ and $\theta_0 = v_0/d$ is compared with experimental data presented by Fukumoto and Kubo (1977) and Nethercot (1978). The theoretical curve provides a satisfactory lower bound solution for the experimental data.

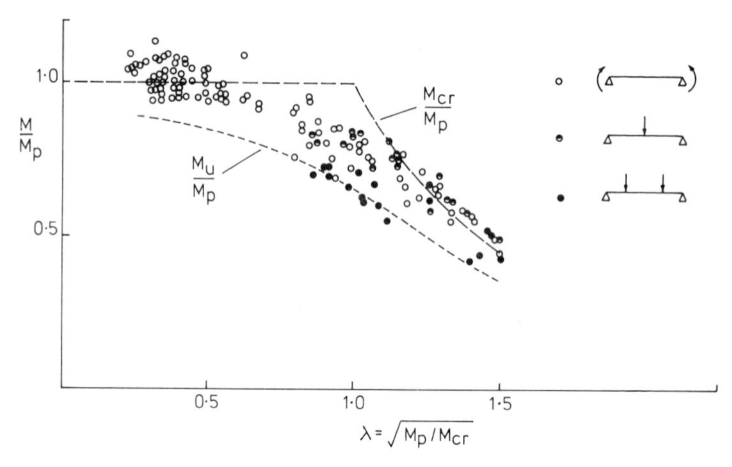

FIG. 5.15. Comparison of theoretical failure envelope with test data.

5.7 SUMMARY AND CONCLUSIONS

It has been shown that the non-linear strains defined in Section 5.2 facilitate the solution of a variety of problems associated with the instability, geometric non-linearity and collapse of thin-walled bars of open cross-section. Closed form solutions can be obtained for the elastic critical loads of beams and columns subjected to simple loading and boundary conditions, but in general, and for all problems of geometric non-linearity, numerical methods have to be used. The theory presented is applicable for all thin-walled bars of open cross-section such as angles, channels, T- and I-sections, having initial geometric imperfections, and subjected to axial, flexural and torsional loading.

REFERENCES

AKAY, H. U., JOHNSON, C. P. and WILL, K. M. (1977) Lateral and local buckling of beams and frames. *J. Struct. Div., Proc. ASCE*, **103**(ST9), 1821–32.

ANDERSON, J. M. and TRAHAIR, N. S. (1972) Stability of monosymmetric beams and cantilevers. *J. Structur. Div., Proc. ASCE*, **98**(ST1), 269–86.

AYRTON, W. E. and PERRY, J. (1886) On struts. *The Engineer*, **62**, 464.

BARSOUM, R. S. and GALLAGHER, R. H. (1970) Finite element analysis of torsional and torsional-flexural stability problems. *Int. J. Numerical Methods in Engineering*, **2**, 335–52.

BLEICH, F. (1952) *Buckling Strength of Metal Structures*, McGraw-Hill, New York.

BREBBIA, C. and CONNOR, J. (1969) Geometrically nonlinear finite element analysis. *J. Eng. Mech. Div., Proc. ASCE*, **95**(EM2), 463–83.

CHAJES, A. (1974) *Principles of Structural Stability Theory*, Prentice-Hall, Englewood Cliffs, N.J.

CHEN, W. F. and ATSUTA, T. (1977) *Theory of Beam Columns*, Vol. 2: *Space Behavior and Design*, McGraw-Hill, New York.

CLARK, J. W. and KNOLL, A. H. (1958) Effect of deflection on lateral buckling strength. *J. Eng. Mech. Div., Proc. ASCE*, **84**(EM2), 1596–1601.

DAVIDSON, J. F. (1952) The elastic stability of bent I-section beams. *Proc. Roy. Soc. London Ser. A*, **212**, 80.

DWIGHT, J. B. (1978) Design of axially loaded columns including interactive buckling. Lecture 3, *Symposium on the Background to the New British Standard for Structural Steelwork*, Imperial College, London.

FUKUMOTO, Y. and KUBO, M. (1977) A survey of tests on lateral buckling strength of beams. *Proceedings, 2nd International Colloquium on Stability of Steel Structures*, ECCS–IABSE–SSRC–CRCJ, Liège.

GALLAGHER, R. H. and PADLOG, J. (1963) Discrete element approach to structural instability analysis. *J. AIAA*, **1**, 1437–9.

JOHNSON, C. P. and WILL, K. M. (1974) Beam buckling by finite element procedure. *J. Struct. Div., Proc. ASCE*, **100**(ST3), 669–85.

KAPUR, K. K. and HARTZ, B. J. (1966) Stability of thin plates using the finite element method. *J. Eng. Mech. Div., Proc. ASCE*, **92**(EM2), 177–95.

KAWAI, T. and YOSHIMURA, N. (1969) Analysis of large deflection of plates by finite element method. *Int. J. Numerical Methods in Engineering*, **1**, 123–33.

LANGHAAR, H. L. (1962) *Energy Methods in Applied Mechanics*, Wiley, New York.

MASUR, E. F. and MILBRADT, K. P. (1957) Collapse strength of redundant beams after buckling. *J. Applied Mechanics, ASME*, **24**(2), 283.

MURRAY, D. W. and WILSON, E. L. (1968) Finite element large deflection analysis of plates. *J. Eng. Mech. Div., Proc. ASCE*, **94**(EM1), 143–65.

NETHERCOT, D. A. (1975) Inelastic buckling of steel beams under non-uniform moment. *Structural Engineer*, **53**(2), 73–8.

NETHERCOT, D. A. (1978) Lateral torsional buckling of beams. Lecture 5, *Symposium on the Background to the New British Standard for Structural Steelwork*, Imperial College, London.

NETHERCOT, D. A. and ROCKEY, K. C. (1971) Finite element solutions for the buckling of columns and beams. *Int. J. Mech. Sci.*, **13**, 945–9.

NETHERCOT, D. A. and ROCKEY, K. C. (1973) Lateral buckling of beams with mixed end conditions. *Structural Engineer*, **51**(4), 133–8.

RAJASEKARAN, S. (1971) Finite element analysis of thin walled members of open section. Ph.D. thesis, University of Alberta, Edmonton.

RAJASEKARAN, S. (1977) Finite element method for plastic beam-columns. Chap. 12 of Chen and Atsuta (1977).

ROBERTS, T. M. (1981) Second order strains and instability of thin walled bars of open cross section. *Int. J. Mech. Sci.*, **23**, 297–306.

ROBERTS, T. M. and ASHWELL, D. G. (1971) The use of finite element mid-increment stiffness matrices in the post-buckling analysis of imperfect structures. *Int. J. Solids Structures*, **7**, 805–23.

ROBERTS, T. M. and AZIZIAN, Z. G. (1982*a*) Influence of pre-buckling displacements on the elastic critical loads of thin walled bars of open cross section. Report, Dept of Civil and Structural Engineering, University College, Cardiff.

ROBERTS, T. M. and AZIZIAN, Z. G. (1982*b*) Nonlinear analysis of flexure and torsion. Report, Dept of Civil and Structural Engineering, University College, Cardiff.

ROBERTS, T. M, and AZIZIAN, Z. G. (1983) Instability of thin walled bars. *J. Eng. Mech. Div., Proc. ASCE*, in press.

ROBERTSON, A. (1925) The strength of struts. Selected Engineering Paper No. 28, Institution of Civil Engineers, London, 55 pp.

SCHMIT, L. A., BOGNOR, F. K. and FOX, R. L. (1968) Finite deflection structural analysis using plate and cylindrical shell discrete elements. *J. AIAA*, **5**, 781–91.

TIMOSHENKO, S. P. and GERE, J. M. (1961) *Theory of Elastic Stability*, McGraw-Hill, New York.

VACHARAJITTIPHAN, P., WOOLCOCK, S. T. and TRAHAIR, N. S. (1974) Effect of in-plane deformation on lateral buckling. *J. Struct. Mech.*, **3**(1), 29–60.

WOOLCOCK, S. T. and TRAHAIR, N. S. (1974) The post-buckling behaviour of determinate beams. *J. Eng. Mech. Div., Proc. ASCE*, **100**(EM2), 151–71.

ZIENKIEWICZ, O. C. (1971) *The Finite Element Method in Engineering Science*, McGraw-Hill, London.

Chapter 6

DIAPHRAGM-BRACED THIN-WALLED CHANNEL AND Z-SECTION BEAMS

TEOMAN PEKÖZ

Department of Structural Engineering, Cornell University, Ithaca, New York, USA

SUMMARY

The behaviour of thin-walled diaphragm-braced beams is discussed and various practical construction details are illustrated. Two types of analytical solutions are presented, one of which is in a form suitable for hand calculations. The solutions have been verified by an experimental programme. Various other considerations and the work in progress are summarised.

6.1 INTRODUCTION

Thin-walled cold-formed steel channel and Z-section flexural members are used widely as floor joists and as purlins in metal building roof systems. These sections are easy and economical to fabricate and erect. However, they are weak in the lateral direction and in torsion. In order to use their full bending capacity in the strong direction, they must be braced in the lateral direction and against twisting. With proper attention to details, the flooring or the roof panels which are connected to these sections do provide to some extent such bracing effect by virtue of their shear rigidity and resistance to local bending at the connections. This chapter deals primarily with channel and Z-section members used as roof purlins. However, the discussion here may be useful in understanding the behaviour of such members in other applications as well. As purlins the flanges of channels are usually provided by a lip stiffener. The Z-sections also have stiffeners. However, the lip stiffeners for Z-sections are in general at less than right-angles with the flanges to facilitate nesting.

Cold-formed steel or aluminium panels and decks are frequently used for roofing or wall sheeting in industrial buildings. Such panels and decks are referred to as diaphragms when they are designed to transfer or support loads through in-plane shear resistance. In addition to their enclosure function, diaphragms provide bracing to the individual purlins and columns, thereby increasing their load-carrying capacity. A typical roof assembly with diaphragm-braced purlins is illustrated in Fig. 6.1. Roof panels are usually thin-walled, corrugated or stiffened orthotropic steel panels.

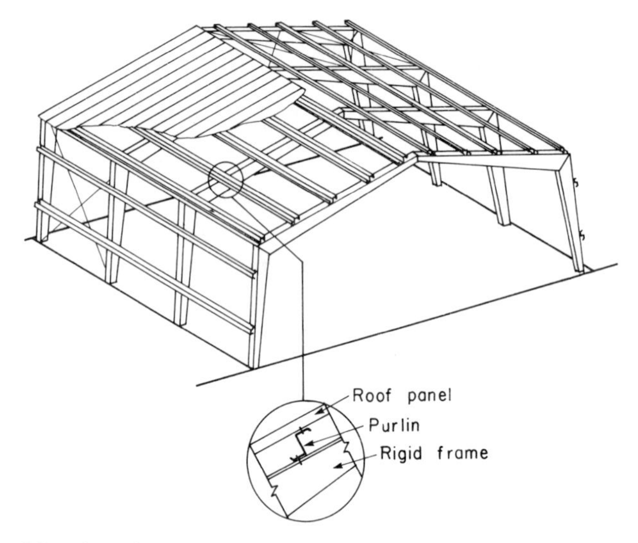

FIG. 6.1. A typical roof assembly with diaphragm-braced purlins.

In the great majority of metal building applications, the purlins are continuous over the building frames which serve as intermediate supports for the purlins. The continuity is accomplished by lapping the purlins over the supports. The Z-section purlins are lapped by nesting one inside the other. Lipped channel purlins are lapped by placing them back-to-back over the supports. In each case the purlins are bolted together and the screws conecting the roof panels to the purlins penetrate through both purlins.

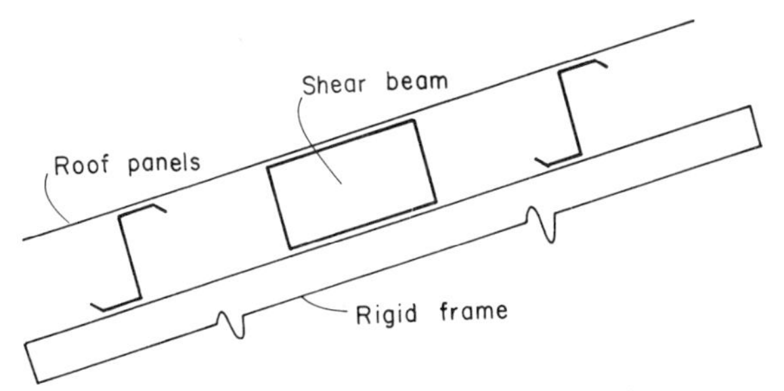

FIG. 6.2. Shear beam (shear channel).

In general, to have an effective restraint by the roof panels, certain conditions must be met. These include:

(a) Proper attachment of the roof panels to the rigid frames and the end walls by some means as shown in Fig. 6.2.
(b) Special considerations are necessary for openings and conditions at the ridge and the eaves to ensure proper support of the panels to carry shear loads.
(c) Proper connection between the purlins and the roof panels. Certain types of connection may result in very ineffective diaphragm bracing.
(d) Proper seam connections between panels.

If the restraint by the roof panels alone is not sufficient, for example, as would be the case if any of the conditions given in the items above are not

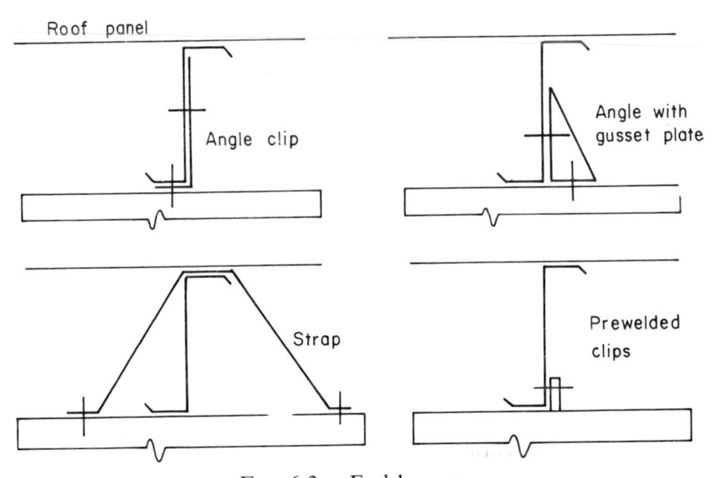

FIG. 6.3. End braces.

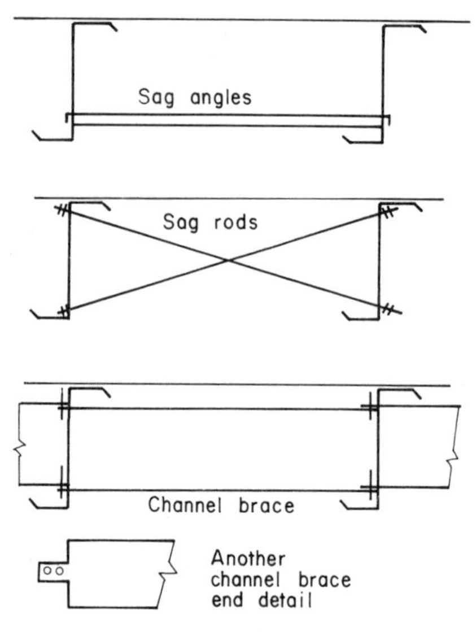

FIG. 6.4. Intermediate braces.

met, then the use of end and/or intermediate braces is one of the ways of enhancing the load-carrying capacity of the purlins.

End bracing is accomplished in general by rigid end clips as shown in Fig. 6.3. In effect, rigid end clips, if they are well designed, may serve as the shear channel or beam as shown in Fig. 6.2 and thus may make the diaphragm action more effective. If full diaphragm action is to be used, then, of course, the other conditions listed must also be satisfied. End bracing also restrains or eliminates purlin roll, and lateral deflection at supports therefore enhances the load-carrying capacity of purlins.

Intermediate braces also take several forms. Depending on the span, one, two or three intermediate braces may be used. A few types of intermediate braces used in practice are shown in Fig. 6.4. As will be discussed below, though the intermediate braces restrain the twisting and the lateral deflection of purlins, they do not always increase the load-carrying capacity.

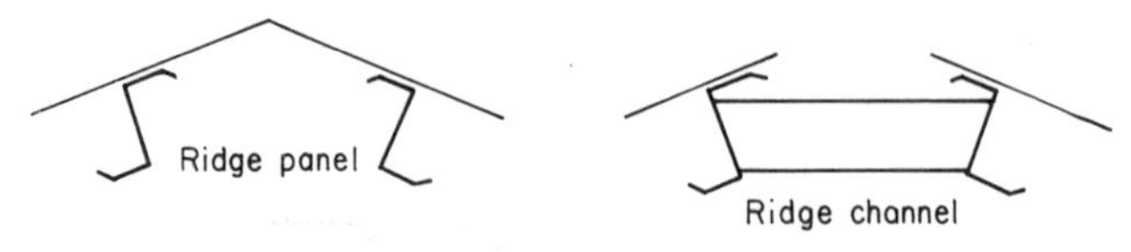

FIG. 6.5. Ridge restraints.

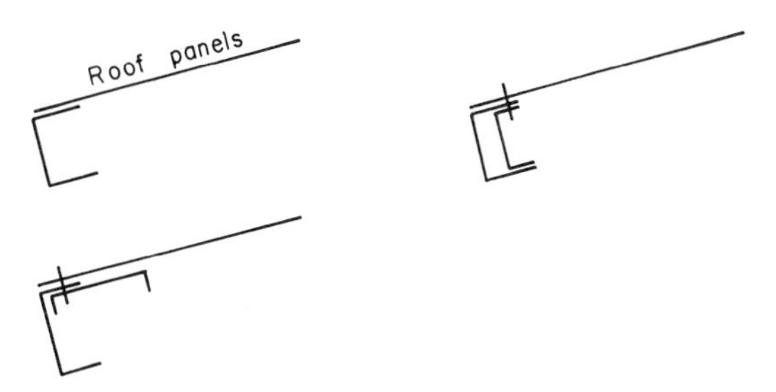

FIG. 6.6. Eave restraints (eave struts).

In addition to the means discussed above, eave or ridge restraints can be used to increase the load-carrying capacity of purlins. These restraints reduce the deflections in the plane of the roof and, hence, reduce the stresses in the purlins. Some typical ridge and eave restraints are shown in Figs. 6.5 and 6.6.

Reversing the purlins as shown in Fig. 6.7 is also used to restrain the lateral deflections and twist. This is done typically for lipped channel purlins and by some companies in some applications for Z-purlins. The problems with modularising are a disadvantage of this type of system.

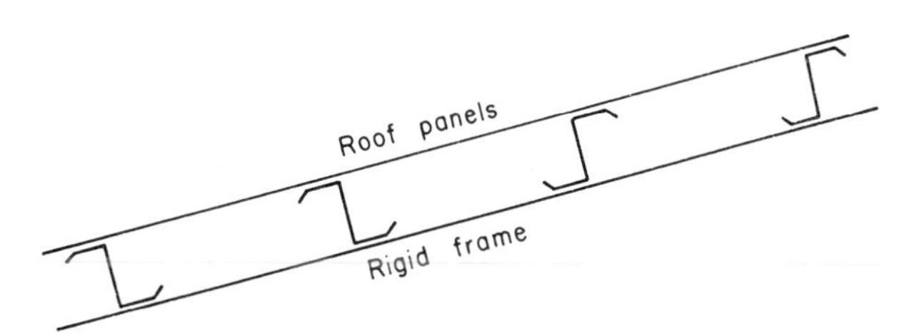

FIG. 6.7. Reversal of purlins.

6.2 ANALYTICAL FORMULATIONS

In this chapter, two approaches to modelling the behaviour will be discussed. The first is based on a modification of the torsional–flexural behaviour theory (Timoshenko and Gere, 1961; Vlasov, 1961). The second is a simplified approach based on the beams on elastic foundation theory. In both approaches it is assumed that the load is in the initial plane of the web, and that the load components perpendicular to the web will be supported by the roof panels. It is essential to provide proper means for anchoring these forces, as discussed above. It is also assumed that the purlins are restrained against 'rolling' (twisting and lateral bending) at the supports.

6.2.1 Torsional–Flexural Behaviour Theory

The bracing action of the panels by virtue of their shear rigidity, w, was first formulated by Larson (1960). Subsequently the general theory, including both the bracing due to shear rigidity of the panels and the restraint offered by the panels to the twisting of beams and columns, was developed by Errera *et al.* (1967), Apparao and Errera (1968) and Apparao *et al.* (1969) for doubly symmetric sections. Later, Celebi (1972) extended the treatment to lipped channel and Z-sections and derived the pertinent differential equations of equilibrium. These differential equations are in the following form:

$$\frac{E(I_x I_y - I_{xy}^2)}{I_x}\frac{\mathrm{d}^4 u}{\mathrm{d}z^4} - Q\frac{\mathrm{d}^2 u}{\mathrm{d}z^2} - Qe_{\mathrm{D}}\frac{\mathrm{d}^2\phi}{\mathrm{d}z^2} + \frac{\mathrm{d}^2}{\mathrm{d}z^2}(M_x\phi) = -\frac{I_{xy}}{I_x}p_y \qquad (6.1)$$

$$EC_{\mathrm{w}}\frac{\mathrm{d}^4\phi}{\mathrm{d}z^4} - GK\frac{\mathrm{d}^2\phi}{\mathrm{d}z^2} - Qe_{\mathrm{D}}\frac{\mathrm{d}^2\phi}{\mathrm{d}z^2} - p_y e\phi + F\phi + M_x\frac{\mathrm{d}^2 u}{\mathrm{d}z^2} - Qe_{\mathrm{D}}\frac{\mathrm{d}^2 u}{\mathrm{d}z^2} = p_y a \qquad (6.2)$$

where u, v are the deflections of the shear centre in the direction of x and y axes, respectively (Fig. 6.8);
ϕ is the twist angle of the cross-section (Fig. 6.8);
I_x, I_y, I_{xy} are the moments of inertia and the product of inertia, respectively;
E, G are the elastic modulus and the shear modulus, respectively;
C_{w}, K are the warping constant and the St Venant torsion constant, respectively;
Q, F are the shear rigidity and the rotational restraint of the diaphragm, respectively;
e_{D} is the vertical distance (positive) of the diaphragm from the shear centre (Fig. 6.9);
a, e are the horizontal and the vertical distances (positive) from the shear centre to the point of application of the load p_y (Fig. 6.9);
p_y is the distributed load in the plane of the web (positive for gravity, negative for uplift);
M_x is the bending moment due to p_y.

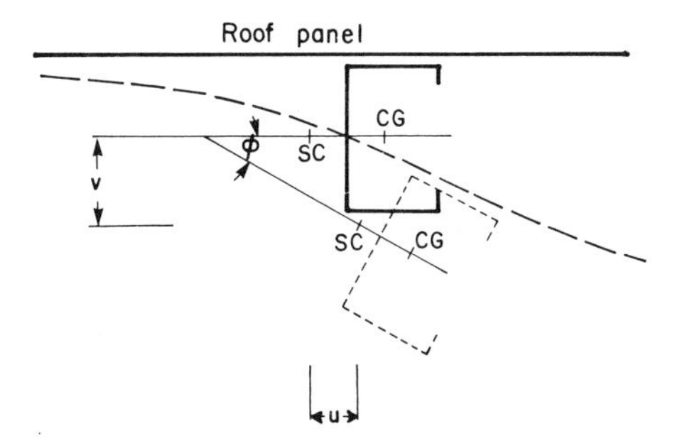

FIG. 6.8. Displacement components.

Celebi (1972) obtained a solution for the differential equations of equilibrium using the Galerkin method. For this solution the displacement components are assumed in the form of series:

$$u = \sum_{n=1}^{\infty} u_n X_n \tag{6.3}$$

$$\phi = \sum_{n=1}^{\infty} \phi_n Z_n \tag{6.4}$$

where X_n, Z_n are the eigenfunctions of nth mode that satisfy the kinematic boundary conditions, and u_n, ϕ_n are the amplitudes for the eigenfunctions

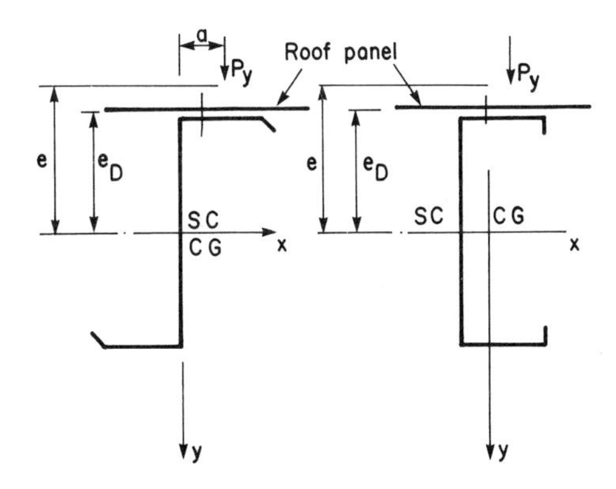

FIG. 6.9. Notation for the point of application of the load.

X_n and Z_n, respectively. The application of the Galerkin method leads to a solution in the following form:

$$\sum_{n=1}^{\infty} \int_0^L u_n \left[\frac{E(I_x I_y - I_{xy})^2}{I_x L^2} X_n'''' - Q X_n'' \right]$$

$$+ \phi_n[- QeZ_n'' + (M_x Z_n)'']X_m \, \mathrm{d}z = -\frac{I_{xy}}{I_x} L^2 \int_0^L p_y X_m \, \mathrm{d}z$$

$$m = 1, 2, \ldots, \infty \qquad (6.5)$$

where the primes indicate differentiation with respect to X, and

$$\sum_{n=1}^{\infty} \int_0^L u_n[M_x X_n'' - QeX_n''] + \phi_n \left[\frac{EC_w}{L^2} Z_n'''' - (GK + Qe^2)Z_n'' \right.$$

$$\left. - p_y e L^2 Z_n + FL^2 Z_n \right] Z_m \, \mathrm{d}z = aL^2 \int_0^L p_y Z_m \, \mathrm{d}z$$

$$m = 1, 2, \ldots, \infty \qquad (6.6)$$

where the primes indicate differentiation with respect to Z. However, it was found (Celebi, 1972) that, in general, three terms of each series need to be included. A detailed discussion of the general behaviour as depicted by this solution is given by Celebi *et al.* (1971).

The integrals in eqns (6.5) and (6.6) were evaluated by Celebi (1972) for the case of simply supported spans. Subsequently, Peköz (1973, 1975), in collaboration with Celebi, obtained solutions for the five additional cases shown in Fig. 6.10. At an end shown as pinned, the conditions $u'' = \phi' = 0$ is assumed, while at a fixed end $u' = \phi = 0$. The rotations are assumed to be fully restrained at intermediate braces. The resulting solutions are too complex for manual computation. Therefore, a computer program (Peköz, 1973) was prepared on the basis of these solutions. Based on the solutions for the cases above, this program can be used in analysing continuous purlins for both gravity and wind uplift conditions.

6.2.2 A Simplified Approach for Wind Uplift Loading

Wind uplift is an important design condition for roof purlins. Under this condition the compression flange is laterally unsupported in large moment regions.

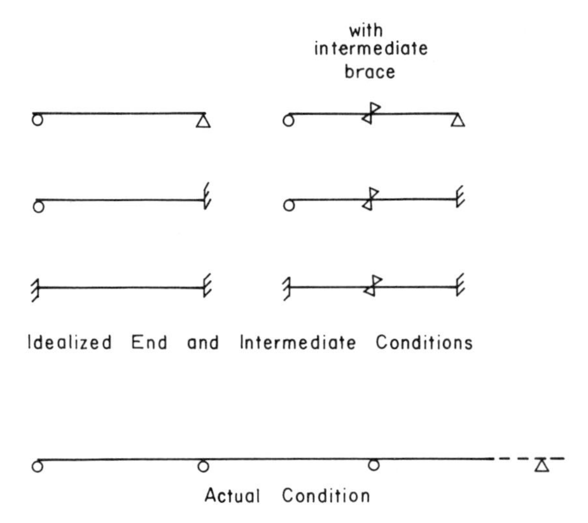

FIG. 6.10. Idealised end and intermediate conditions.

(*a*) *Stresses and Displacements*
Because of its complexity the classical theory of torsional flexure is not suitable for treating the effects of local buckling and post-buckling behaviour on the overall behaviour. Furthermore, this approach cannot be extended easily to include the effects of initial sweep and twist of purlins. The importance of these parameters was observed in several large-scale tests. The torsional–flexural behaviour theory was shown to be satisfactory for predicting deflections but not ultimate loads. The discrepancies were larger for thinner sections, indicating the importance of local behaviour of the component plate elements of the sections.

In the design of purlins the approach of Section 3 of Part III of AISI (1977) is used quite frequently. This approach also has several deficiencies. First, it is assumed that the compression flange of a purlin does not deflect laterally until failure. The actual behaviour is clearly otherwise: the compression flange deflects laterally from the start of loading. Second, the effect of initial sweep and twist is not accounted for. These and several other additional deficiencies of that approach have been eliminated by a new approach derived by Peköz and Soroushian (1982*a*). Large-scale and component tests have shown that the approach is satisfactory for design purposes. The discussion of this approach is taken mainly from Peköz and Soroushian (1982*b*).

In order to reach a simple solution, and based on an intuitive assessment of the overall behaviour in the building, the purlins are assumed fixed against deflection in the plane of the roof at the purlin to roof panel connection. This is equivalent to taking the value of the shear rigidity to be infinite in the method of analysis of AISI (1968), Celebi *et al.* (1971) and Peköz (1975). This assumption appears reasonable on the basis of studies conducted using the computer program of Peköz (1975). These studies show that, as long as the value of the shear rigidity is larger than a reasonable minimum, this assumption is satisfactory for the simple solutions obtained. These solutions were obtained for simply supported purlins. However, they can be readily extended to continuous purlins as discussed below.

The lipped channels and Z-sections undergo vertical deflections and twisting. The twisting results in lateral deflections of the compression flange. Deflected configurations are shown in Fig. 6.11(a). For the purposes of our discussion and the derivation of the analytical model, the deformations can be considered in two stages. These stages will be referred to as the torsion and the vertical bending stages. These stages are illustrated in Fig. 6.11(b).

The vertical bending stage can be analysed using the simple flexure theory. However, the moment of inertia is to be computed for the twisted section. Twisting does introduce some small vertical deflection component which will be added to the vertical deflections obtained for the vertical bending stage.

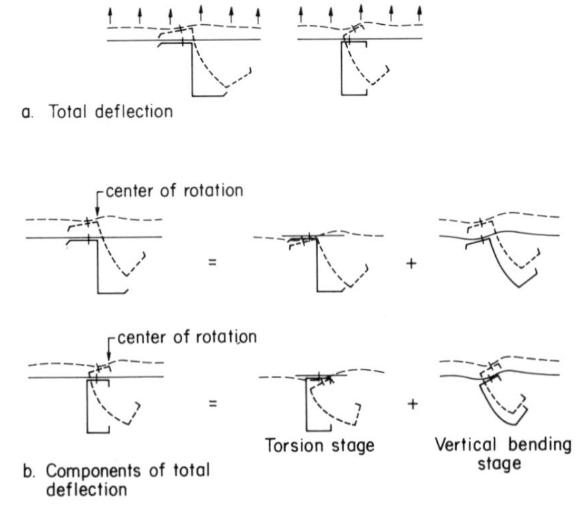

FIG. 6.11. Idealised behaviour of purlins.

The torsion stage involves lateral deflection and twisting which will be analysed through the use of an idealised analytical model. The model involves the assumption of a beam column on elastic foundation. The beam column section consists of the compression flange and a portion of the web drawn with thick lines in Fig. 6.13. The spring constraint for the elastic foundation is obtained as follows. The purlin to panel connection can be idealised to act as a rotational spring located at the centre of rotation of each purlin as shown in Fig. 6.12(a). Further simplification is made by converting this spring into a linear extensional spring of stiffness k located at the level of the compression flange shown in Fig. 6.12(a). This linear spring combines the effect of the restraint provided by the roof panels and the web of the purlin to the compression portion of the purlin. The roof panel restraint is best determined by test.

The lateral force on the idealised beam column results from the variation of the shear flow along the member. In the case of Z-sections, the centre of rotation can be assumed to be the corner between the web and the tension flange, and the flange shear flow force resultant causes a twisting moment about the centre of rotation. The shear flow force in the web goes through the centre of rotation and hence causes no twisting moment. On the other hand, the centre of rotation for the lipped channels can be assumed as the junction of the tension flange and its stiffener. In this case, the shear flow force in the web does not go through the centre of rotation. Thus the shear force in the web, in addition to the shear force in the flange, causes a

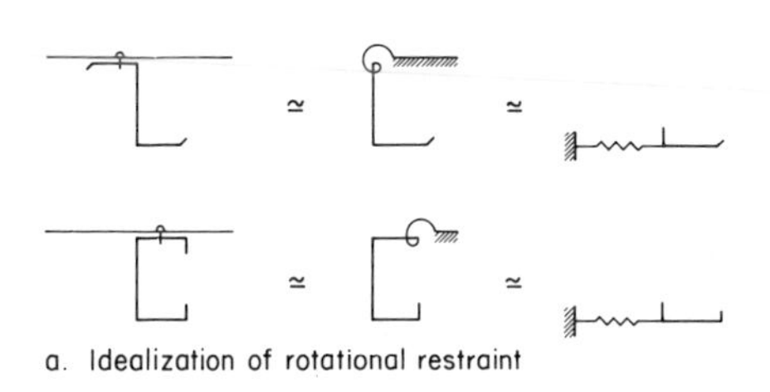

a. Idealization of rotational restraint

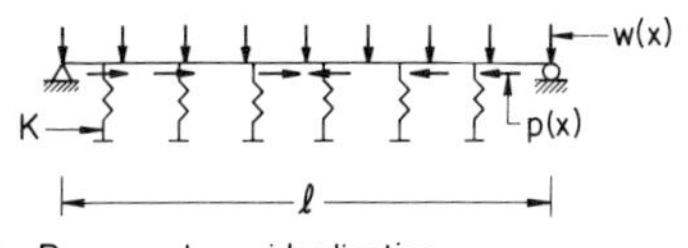

b. Beam column idealization

FIG. 6.12. Behaviour idealisation.

twisting moment about the centre of rotation. The change in the shear force in the web per unit length is equal to the applied uniform vertical load q.

The distributed lateral load $w(x)$ on the idealised beam column (Fig. 6.12(b)) results from the differences in the shear flow forces along the length of the member and can be expressed as

$$w(x) = \frac{\text{Flange shear force at } (x + \mathrm{d}x) - \text{Flange shear force at } (x)}{\mathrm{d}x} + \alpha q \tag{6.7}$$

$$w(x) = q\left(\frac{\bar{Q}b}{2I} + \alpha\right) \tag{6.8}$$

where q = distributed uplift load on purlin;
b = flange width (see Fig. 6.13);
$\bar{Q}$ = static moment of the flange and the stiffening lip around the centroidal axis of the purlin;
I = moment of inertia about the horizontal axis of the section in the deflected configuration. Either the entire gross or the effective section is to be used as applicable. The determination of I will be discussed below and defined in eqn (6.20);
α = the distance from the centre of rotation to the junction of flange and web divided by the depth; it is 0 for Z-section purlins.

The first term of the right-hand side of eqn (6.7) represents the lateral loading on the idealised beam column due to the shear flow in the

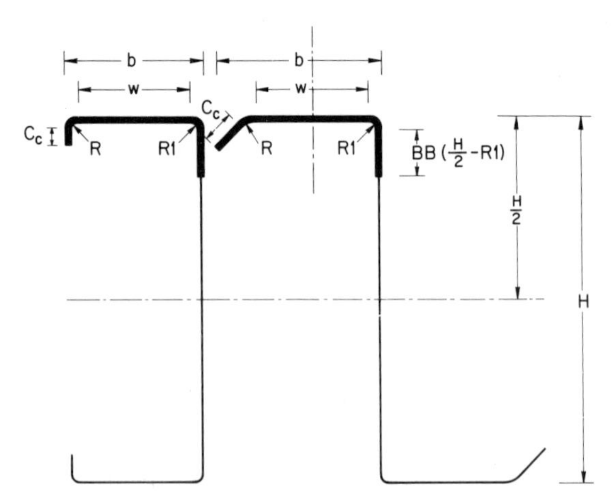

FIG. 6.13. Cross-sectional dimensions. The heavy lines indicate the idealised beam column section.

compression flange and is the same for both the lipped channel and the Z-section purlins. For the sake of simplicity, the flange is assumed to be a flat element of width b (see Fig. 6.13). This simplification is used only in determining $w(x)$ and $p(x)$ derived below. The second term of the same equation is to account for the twisting moment due to the shear force in the web. To facilitate the derivation below, the loading in the web is converted to a load applied at the flange level and which results in the same twisting moment around the centre of rotation as the load in the web.

The distributed axial force $p(x)$ results from the variation of the compression stress along the length of the member. The beam column section is assumed to be the flange and part of the flat width of the compression portion of the web $BB(H/2 - R_1)$ as shown in Fig. 6.13. The axial force can be expressed as follows:

$$p(x) = \frac{\text{Beam column axial force at } (x + \mathrm{d}x) - \text{Beam column axial force at } (x)}{\mathrm{d}x} \tag{6.9}$$

which is

$$p(x) = V \, . \, G \tag{6.10}$$

where G = static moment of the beam column area about the centroidal axis of the purlin divided by the moment of inertia;
V = shear force at any point in the span (function of x) = $\mathrm{d}M/\mathrm{d}x$.

With the above idealisations the analysis for lateral deflections and twisting (torsion stage) is reduced to the analysis of the beam column section shown in Fig. 6.13 with the loading shown in Fig. 6.12(b).

The strain energy consists of two parts: the flexural strain energy (U_f) and the elastic foundation strain energy (U_k). Their values are:

$$U_\mathrm{f} = 2 \int_0^{l/2} \frac{EI_\mathrm{f}}{2} \left(\frac{\mathrm{d}^2 u}{\mathrm{d}x^2} - \frac{\mathrm{d}^2 u_0}{\mathrm{d}x^2} \right)^2 \mathrm{d}x \tag{6.11}$$

$$U_\mathrm{k} = 2 \int_0^{l/2} \frac{K(u - u_0)^2}{2} \, \mathrm{d}x \tag{6.12}$$

where I_f = moment of inertia of the beam column around its own centroidal axis parallel to the web;
l = span of the beam column;
K = stiffness of the linear spring;
u = deflection of the beam column in the plane of the flange;
u_0 = initial sweep of the beam column in the plane of the flange;
E = modulus of elasticity.

The potential energy of the external loads also has two parts: that of the lateral loads (U_w) and the axial loads (U_p). Their values are:

$$U_w = -2 \int_0^{l/2} w(x) . (u - u_0) \, dx \tag{6.13}$$

$$U_p = -2 \int_0^{l/2} \left\{ p(x) \cdot \left(\frac{1}{2}\right) \int_x^{l/2} \left[\left(\frac{du}{dx}\right)^2 - \left(\frac{du_0}{dx}\right)^2 \right] dx \right\} dx \tag{6.14}$$

The total potential energy can be expressed as

$$U = V_f + V_k + W_w + W_p \tag{6.15}$$

The following displacement functions satisfy the end conditions for the simply supported purlins:

$$u = \sum_{n=1,3,\ldots} a_n \sin \frac{n\pi x}{l} \tag{6.16}$$

$$u_0 = \sum_{n=1,3,\ldots} a_{n0} \sin \frac{n\pi x}{l} \tag{6.17}$$

where a_n are the amplitudes of deflections to be computed using the energy expressions and a_{n0} are the amplitudes which can be obtained from the measured values of the sweep in the plane of the flange. Since the lateral deflections are symmetric with respect to the mid-span, only the odd values of n are used in the solution. Solutions for various intermediate bracing conditions are given by Apparao and Errera (1968). Here only the simply supported case will be discussed.

The amplitudes a_n can be determined by the Ritz procedure as:

$$a_n = \frac{4ql^4 \left(\dfrac{\bar{Q}b}{2I} + \alpha\right) + a_{n0}(n^5 EI_f \pi^5 + nKl^4\pi)}{EI_f\pi^5 n^5 + Kl^4 n\pi - 4Gql^4 n\pi(0{\cdot}206n^2 - 0{\cdot}063)} \tag{6.18}$$

It should be noted that the lateral deflection, u, is the total lateral deflection of the compression flange including the initial sweep. The deflection of the vertical bending stage does not have a component in the plane of the flange. However, the lateral deflection, u, causes a deflection component in the vertical direction. The vertical deflection component is

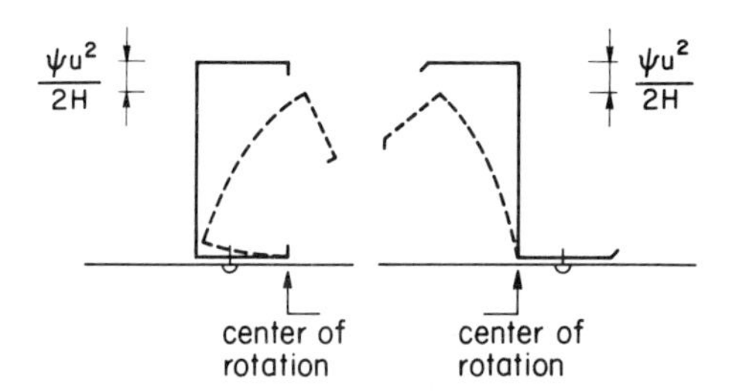

FIG. 6.14. Vertical deflection resulting from lateral deflection and twist.

shown in Fig. 6.14. The total vertical deflection, v, can be expressed as follows:

$$v = \frac{qx}{24EI}(l^3 - 2lx^2 + x^3) + \frac{\psi u^2}{2H} \tag{6.19}$$

where ψ is a correction factor which will be discussed below, and H is the height of the section.

As a result of lateral deflections and twist, the moment of inertia with respect to the centroidal axis perpendicular to the original position of the web will be reduced. The following simple approximate expression for the reduced moment of inertia is obtained from Peköz and Soroushian (1982*a*):

$$I = I_0\left[1 - \left(\frac{u}{H}\right)^2\right] \tag{6.20}$$

where I_0 is the moment of inertia of the total or effective section about the centroidal axis perpendicular to the web.

The total stresses in the purlin are obtained by superposing the stresses due to bending in the plane of the web (the vertical bending stage) on the stresses due to twisting and lateral deflection (torsion stage). Thus the total stress can be expressed as

$$\sigma = \frac{M}{S} + \frac{M_f}{S_f} \tag{6.21}$$

where M = moment resulting from the vertical bending stage $= q(1 - x)/2$;
S = section modulus based on the moment of inertia I;
S_f = section modulus of the beam column about its centroidal axis parallel to the web;
M_f = beam column bending moment.

The beam column bending moment M_f can be determined as follows:

$$M_f = EI_f(u'' - u_0'') \tag{6.22}$$

$$M_f = \frac{EI_f \pi^2}{l^2} \cdot \sum_{n=1,3,\ldots} n^2(a_n - a_{n0}) \sin \frac{n\pi x}{l} \tag{6.23}$$

The maximum compressive stress occurs at the junction of the compression flange with the web. Thus, S_f to be used in eqn (6.19) should be the appropriate section modulus for that point.

Numerical studies were conducted using practical values of the parameters involved. These studies have shown that the series used in the Ritz procedure solution converges rapidly. Taking only one term of the series leads to errors in u of less than 5%. For $n = 1$, a_n becomes

$$a_1 = \frac{C_1\left(\dfrac{\bar{Q}b}{2I} + \alpha\right) + a_{10}}{1 - 0{\cdot}45C_1G} \tag{6.24}$$

where

$$C_1 = \frac{1{\cdot}27q}{\dfrac{94{\cdot}41EI_f}{l^4} + K} \tag{6.25}$$

The flange bending moment can now be found using eqn (6.23). Because of the term n^2 in the summation, taking only the term with $n = 1$ leads to less accurate results, namely, the convergence is less rapid for the flange bending moment than that for the lateral deflection. However, with proper simplifications, as will be described below, taking $n = 1$ even for the flange bending moment gives excellent results.

Based on some parametric studies and the evaluation of the test results, the following simplifications were made. If the portion of the web contributing to I_f, Q and G is ignored and the compression flange width is taken to be equal to b as shown in Fig. 6.13, then

$$I_f = tb^3/12 \tag{6.26}$$

$$G = btH/4I_0 \tag{6.27}$$

and

$$\bar{Q} = btH/2 \tag{6.28}$$

Substituting eqns (6.26)–(6.28) into eqns (6.24) and (6.25), the following is obtained:

$$a = \frac{C(Zb + \alpha) + a_0}{1 - 0{\cdot}9ZC} \tag{6.29}$$

where

$$Z = \frac{tHb}{4I_0} \tag{6.30}$$

$$C = \frac{1{\cdot}27q}{\dfrac{7{\cdot}87Etb^3}{l^4} + K} \tag{6.31}$$

α was defined in connection with eqn (6.8). All parameters should have consistent units.

In the above equations the notation has been simplified by expression C_1 as C, a_1 as a, and a_{10} as a_0. Furthermore, for simplicity, the reduction in the moment of inertia is ignored. Using eqn (6.21) and the simplifications discussed above, the expression for the maximum stress at the flange to web junction for Z-purlins and flange to stiffening lip junction of C-purlins becomes

$$\sigma = \frac{MH}{2I} + \frac{Eb\pi^2}{2l^2}(a - a_0) \tag{6.32}$$

M is defined in connection with eqn (6.21), I is defined by eqn (6.20) with u = the amount of lateral deflection for lateral load q. The following procedure for determining the value of K in the above equations gave good results. The horizontal force, $w(x)$, at which the value of K is determined can be found by eqn (6.8). With the simplifications above, this equation becomes

$$w = q\left(\frac{b^2tH}{4I_0} + \alpha\right) \tag{6.33}$$

Units of w are in lb in^{-1}.

The solution involves a very rapid converging iteration. In this iteration, first a failure load q is assumed. Then the value of w is determined. From a plot of lateral load vs. lateral displacement of an F-test (described below), the value of K is determined at a load equal to w.

The total maximum vertical deflection can be obtained from eqn (6.19) by setting $x = l/2$:

$$v_{max} = \frac{5ql^3}{384EI} + \frac{\psi a^2}{2H} \tag{6.34}$$

The first term in the above equation is the component due to vertical bending. $a^2/2H$ in the second term is due to lateral bending as shown in Fig. 6.14. The factor ψ is to account for the cross-sectional distortion effects. It should depend on the cross-sectional dimensions. However, a regression analysis (Peköz and Soroushian, 1982*a*) conducted on the test results indicated that taking a value of 3·4 for ψ leads to good agreement between the computed and observed results.

(*b*) *Failure Criteria*

Under uplift loading, simply supported purlins deflect as shown in Fig. 6.11(a). Thus, in general the maximum compressive stress occurs at the junction of the web to compression flange. With only a few exceptions, the dominant failure mechanism observed in the large-scale tests was the formation of an inelastic local buckle at the web to compression flange junction. For this reason, various procedures of predicting web failure were used in the research to formulate a purlin criterion. However, here only the procedure adopted in the AISI (1980) Specification will be discussed. In this approach (based on LaBoube and Yu, 1978) a failure stress is defined. The web is taken as fully effective and the beam column section is assumed to be the compression portion of the section with BB (see Fig. 6.3) taken equal to zero.

If the compression flange is a stiffened plate element, the failure stress, F_{wu}, can be taken as

$$F_{wu} = \left[1{\cdot}210 - 0{\cdot}000\,337\left(\frac{H}{t}\right)\cdot\sqrt{F_y}\right]\cdot F_y \leq F_y \tag{6.35}$$

where F_y is the yield stress of the material.

If the compression flange is an unstiffened plate element, the F_{wu} becomes

$$F_{wu} = \left[1{\cdot}259 - 0{\cdot}000\,508\left(\frac{H}{t}\right)\cdot\sqrt{F_y}\right]\cdot F_y \leq F_y \tag{6.36}$$

In both cases F_{wu} is the stress in the flange at failure.

For the Z-purlins with inclined stiffeners, the web failure stress equation for unstiffened flanges was used in all cases. For the lipped channel purlins, the decision as to whether the flange is stiffened or not can be based not on the AISI Specification but on the procedure proposed by Desmond *et al.* (1981) as follows:

Define

$$R = \frac{w}{t} \qquad R_\alpha = \frac{221}{\sqrt{F_y}} \qquad R_\beta = \frac{77{\cdot}23}{\sqrt{F_y}} \tag{6.37}$$

where w = flat width of flange and t = thickness.

If $R \leq R_\beta$,

$$I_{s_{\text{adeq.}}} = 0 \tag{6.38}$$

If $R_\beta \leq F \leq R_\alpha$,

$$I_{s_{\text{adeq.}}} = 389t^4 \left(\frac{R}{R_\alpha} - 0{\cdot}324\right)^3 \tag{6.39}$$

If $R \geq R_\alpha$,

$$I_{s_{\text{adeq.}}} = t^4 \left(\frac{115R}{R_\alpha} + 5\right) \tag{6.40}$$

If $I_s > I_{s_{\text{adeq.}}}$, assume the flanges to be unstiffened.

If $I_s \leq I_{s_{\text{adeq.}}}$, assume the flange to be stiffened.

As mentioned above, the dominant mode of failure in the tests was due to yielding or local buckling at the flange to web junction. In a roof system, under certain circumstances, web crippling may occur. Therefore, this type of failure also needs to be checked separately.

The compression flanges in all the tests were fully effective according to the findings of Desmond *et al.* (1981). If the compression flange is not fully effective, the simple design approach above needs to be modified by taking the effective portions of the flange in eqns (6.26)–(6.31).

Further complications may arise if the initial purlin imperfection is such that maximum stress may occur at the flange tip. In this case the failure criteria above would have to be modified. Further studies on this subject are being conducted at the present time.

6.3 BRACING ACTION PARAMETERS

The parameters that pertain to the bracing action of the diaphragm include shear rigidity Q and rotational restraint F. Though significant advances have been made in computing Q (AISI, 1968; Celebi, 1972; Bryan, 1973;

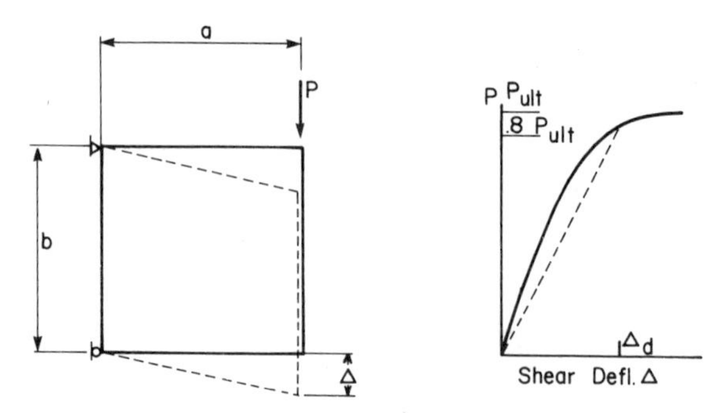

FIG. 6.15. Determination of shear rigidity Q.

ECCS, 1977), in the USA the current general practice is to determine it by a cantilever shear test. The procedure and the test set-up are described in AISI (1968). Figure 6.15 gives a schematic of the test and a typical load deflection plot. A reliable design value for the effective shear modulus, G'_{dr}, for the diaphragm in question can be determined on the basis of the load–deflection plot as follows (Apparao and Errera, 1968):

$$G'_{dr} = \frac{2}{3}\left[\frac{0{\cdot}8P_{ult}a}{\Delta_d b}\right] \tag{6.41}$$

where P_{ult} is the ultimate load in the cantilever shear test;
Δ_d is the deflection corresponding to a load of $0{\cdot}8P_{ult}$;
a, b are the dimensions as shown in Fig. 6.15;
$\frac{2}{3}$ is a factor to obtain a reliable value for shear modulus for design.

A reliable design value for Q is obtained using G'_{dr} as:

$$Q_{dr} = G'_{dr} \,.\, w \tag{6.42}$$

where w is the width of the diaphragm contributing to the bracing of one purlin.

The rotational restraint, F, and the spring constant, K, are also determined experimentally. These parameters are functions of both the cross-bending rigidity of the diaphragm, the local rigidity of the diaphragm to resist the twisting of the purlin, and the rigidity of the purlin section. The cross-bending rigidity is much higher than the local rigidity. Thus, the local rigidity determines the rotational restraint. For the case of uplift loading, the rotational restraint can be found using a test set-up as shown in Fig. 6.16.

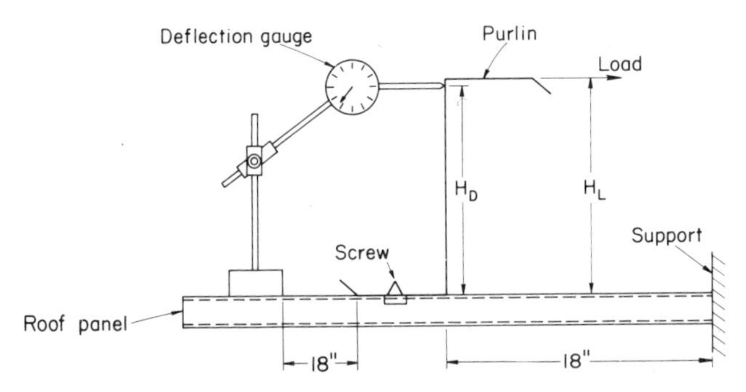

FIG. 6.16. Rotational restraint test.

The rotational restraint is defined as

$$F = \frac{M}{\theta} \tag{6.43}$$

where M is the torsional moment per unit length of the purlin and θ is the rotational angle of the purlin.

The spring constant K is defined as force required per unit length of purlin divided by the deflection of the flange where the force is applied. The determination of K is discussed in more detail in the preceding section.

6.4 EXPERIMENTS

A rather extensive experimental programme involving more than 20 tests was undertaken at Cornell University to check the validity of the several assumptions and simplifications made in the development of the analytical model. An overview of these tests conducted under the supervision of the author is given by Apparao and Errera (1968), Peköz (1975), Razak and Peköz (1980) and Peköz and Soroushian (1982*a*). Other tests have been and are being conducted by various metal building manufacturers. The sections used in these tests are typical of those used in the metal building industry. The Z-sections are 8–9·5 in deep with flange widths between 2 and 2·2 in and thicknesses between 0·06 and 0·12 in. The lipped channels are 0·08 in thick, 7–9 in deep and with flange widths between 1·67 and 1·82 in. The test results show that the formulations discussed above are quite satisfactory as long as the lateral deflections are not more than one-fourth of the section depth (Celebi *et al.*, 1971; AISI, 1977). The deflection prediction formulae are satisfactory up to deflections one-third of the section depth.

6.5 COMPARISON OF THE BEHAVIOUR OF LIPPED CHANNELS AND Z-SECTIONS

Based on numerical studies using both methods of analysis as well as on a limited number of tests, it is concluded that the lipped channel sections deflect laterally 2–3 times more than the Z-sections of the same depth and perimeter. The ultimate loads for lipped channels are underestimated significantly if the lateral deflections are more than about one-fourth the depth of the section. However, the deflections are predicted rather accurately. It would be prudent to establish a maximum allowable lateral deflection criterion for design purposes.

The primary reason for the difference in the basic behaviour can be explained physically as follows. When a lipped channel is loaded with a lateral load parallel to the web and passing through the centroid, it has a strong tendency to twist. The twist is best resisted by the rotational restraint of the roof panels at the screw connections. On the other hand, when the Z-section is subjected to a similar load, it has a strong tendency to deflect laterally. The roof panels having proper construction details as discussed at the beginning of the chapter are more effective in resisting lateral deflections than rotations.

6.6 CONCLUDING REMARKS

Various considerations and formulations for the analysis and design of diaphragm-braced lipped channel and Z-section purlins have been discussed above.

Work is presently underway at Cornell University to extend the approach used for the simple solution discussed above to the case of gravity loading. In the case of gravity loading, taking proper account of the roof shear rigidity and the behaviour of stiffening lips is essential. The latter point is due to the fact that under gravity loading the lateral bending stress adds to the vertical bending stress at the stiffening lip. In the case of uplift loading, the lateral bending stress reduces the total stress at the stiffening lip.

Simple solutions were obtained by Peköz and Soroushian (1982*a*) for wind uplift for the case of one or two intermediate braces. Tests are being conducted by a metal building manufacturer in the USA to verify the approaches developed. The preliminary results show that, as predicted analytically, the ultimate load-carrying capacity is not increased and in some cases is decreased owing to the presence of the intermediate braces. Such braces force the maximum stress to occur at the stiffener of the section rather than at the web–flange corner, thus leading to lower load-carrying capacity. Research on failure criteria for such cases is being planned at Cornell University.

An experimental programme is being carried out by a European metal building manufacturer to compare the behaviour of lipped channel and Z-sections of similar dimensions.

REFERENCES

AISI (1968) *Design of Light Gage Steel Diaphragms*, American Iron and Steel Institute, New York.

AISI (1977) *Cold-Formed Steel Design Manual*, American Iron and Steel Institute, Washington, DC.

AISI (1980) *Specification for the Design of Cold-Formed Steel Structural Members*, American Iron and Steel Institute, Washington, DC (3 Sept.).

APPARAO, T. V. S. R. and ERRERA, S. J. (1968) Design recommendations for diaphragm-braced beams, columns and wall studs. Report No. 332, Dept of Structural Engineering, Cornell University, Ithaca, N.Y.

APPARAO, T. V. S. R., ERRERA, S. J. and FISHER, G. P. (1969) Columns braced by girts and a diaphragm. *Journal of the Structural Division, Proc. ASCE*, **95**, Paper No. 6576, 965–90.

BRYAN, E. F. (1973) *The Stressed Skin Design in Steel Buildings*, Crosby Lockwood Staples, London.

CELEBI, N. (1972) Behavior of channel and Z-section beams braced by diaphragms. Report No. 344, Dept of Structural Engineering, Cornell University, Ithaca, N.Y.

CELEBI, N., PEKÖZ, T. and WINTER, G. (1971) Behavior of channel and Z-section beams braced by diaphragms. *Proceedings, 1st Specialty Conference on Cold-Formed Steel Structures*, University of Missouri, Rolla.

DESMOND, T. P., PEKÖZ, T. and WINTER, G. (1981) Edge stiffeners for thin-walled members. *Journal of the Structural Division, Proc. ASCE*, **107**(ST2), Proc. Paper 16056, 329–51.

ECCS (1977) *European Recommendations for Stressed Skin Design of Steel Structures*, ECCS Committee 17, Constrado, UK.

ERRERA, S. J., PINCUS, G. and FISHER, G. P. (1967) Columns and beams braced by diaphragms. *Journal of the Structural Division, Proc. ASCE*, **93**, Paper No. 5103, 295–318.

LABOUBE, R. A. and YU, W.-W. (1978) Webs for cold formed steel flexural members. Civil Engineering Study 78-1, Structural Series, University of Missouri, Rolla, Final Report (June).

LARSON, M. A. (1960) Discussion of 'Lateral bracing of columns and beams'. *Trans. ASCE*, **125**.

PEKÖZ, T. (1973) Diaphragm braced channel and Z-section purlins. Report and computer program prepared for MBMA and AISI.

PEKÖZ, T. (1975) Progress report on cold-formed steel purlin design. *Proceedings, 3rd International Conference on Cold-Formed Steel Structures*, Dept of Civil Engineering, University of Missouri, Rolla (Nov.)

PEKÖZ, T. and SOROUSHIAN, P. (1982*a*) Behaviour of C- and Z-purlins under wind uplift. Report No. 81-2, Dept of Structural Engineering, Cornell University, Ithaca, N.Y. (Feb.).

PEKÖZ, T. and SOROUSHIAN, P. (1982*b*) Behavior of C- and Z-purlins under wind uplift. *Proceedings, 6th International Conference on Cold-Formed Steel Structures*, St. Louis, Mo. (Nov.)

RAZAK, M. A. A. and PEKÖZ, T. (1980) Progress report: ultimate strength of cold-formed steel Z-purlins. Report No. 80-3, Dept of Structural Engineering, Cornell University, Ithaca, N.Y. (Feb.).

TIMOSHENKO, S. P. and GERE, J. M. (1961) *Theory of Elastic Stability*, McGraw-Hill, New York.

VLASOV, V. Z. (1961) *Thin-Walled Elastic Beams*, 2nd Edn, National Science Foundation, Washington, DC.

Chapter 7

DESIGN OF BEAMS AND BEAM COLUMNS

GEORGE C. LEE and N. T. TSENG

Faculty of Engineering and Applied Sciences,
University of New York at Buffalo, New York, USA

SUMMARY

This chapter discusses some of the rational bases for the proportioning of beam columns with given bending moments and axial force. Both the allowable stress design and the limit design are considered. The AISC Specifications for beam column design are reviewed as a typical example because they are widely used and are conceptually simple. Although the linear interaction equations of the AISC type are derived originally for members in a braced frame subjected to in-plane bending only, they have been extended to cover most other failure modes of beam columns. It is because of the latter that this design practice contains certain undesirable features which are discussed. These are: (1) the factors of safety for different failure modes; (2) the P–Δ effect for unbraced frames; (3) the effect of rigid floor diaphragm action on the effective length factor; (4) the end restraints on lateral–torsional buckling; and (5) the non-linear nature of the interaction equation. Certain alternative approaches and available information for beam column design are also briefly reviewed.

7.1 INTRODUCTION

The behaviour and design of beam columns has been a major research subject in structural engineering for the past several decades. Through an enormous amount of investigations, both experimental and analytical, a clear understanding of the static strength and behaviour of beam columns has been achieved (see, for example, SSRC, 1976; Chen and Atsuta, 1976, 1977; Chen and Cheong-Siat-Moy, 1980). Recent research has given increasing attention to the effect of end restraints, non-proportional loadings and dynamic responses on the behaviour of beam columns, particularly those assocated with non-linear (both geometrical and material) characteristics.

Nearly all members in a typical structure are beam columns. They may respond in a variety of fashions to the imposed loads on the frame, depending upon the geometry of the cross-section, the end restraints, the residual stresses, initial imperfections and the material properties. Failure of beam columns may generally be described by three different modes: (1) local buckling; (2) overall in-plane failure due to excessive bending; (3) out-of-plane lateral–torsional buckling. Local buckling is an important design criterion and is not discussed in this chapter.

Beam columns subjected to bending moment about the strong axis of the cross-section, but laterally unsupported, are commonly used. If such a column is relatively short or torsionally strong (such as the box section), it will typically have in-plane failure. For relatively long and torsionally weak beam columns of open cross-section that are laterally unsupported, lateral–torsional buckling usually governs the design. The typical failure modes for a wide-flange beam column are shown schematically in Fig. 7.1.

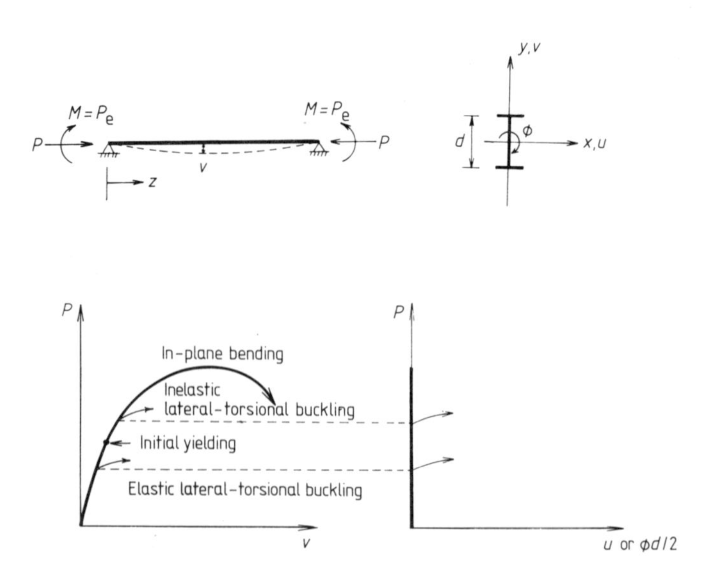

FIG. 7.1. Effect of lateral–torsional buckling on the strength of beam columns.

Most beam columns do not occur as isolated, simply supported members, but as integral parts of a frame which often has rigid connections. In such cases a rational design procedure should take into account the end restraints afforded by the adjacent framing members. The ideal situation for the design of beam columns is to consider iteratively their role in the strength of the entire structure, which is presently not feasible. This will require a substantial change in the design procedure and philosophy in structural engineering practice. Perhaps in some distant future days, when computer technology becomes more sophisticated and simple, such a change will take place. For the moment, the traditional approach in structural design will be assumed in subsequent discussions, that is, an idealised two-dimensional structure is first analysed and free bodies of the beam columns are isolated. They will be checked against various instability failures for the assumed cross-sections with known forces and moments.

The fundamental issue of beam column design is to interpret appropriately the interaction between the axial force and the bending moments that are acting simultaneously on the member. The two extreme conditions are a centrally loaded column and a beam, consistent with the prescribed support conditions along the member and at its ends. Axially loaded columns have been discussed in Chapter 1 of *Axially Compressed Structures* and laterally unsupported beams have been discussed in Chapter 1 of this book. They will not be repeated in this chapter. Also, the mathematical derivations of the equilibrium equations and the solutions for the beam columns subjected to various loading and boundary conditions will not be discussed in this chapter. This information may be found, for example, in the books of SSRC (1976) and Chen and Atsuta (1976, 1977). This chapter will discuss some of the rational bases for the design of steel beam columns with thin-walled open cross-sections. The behaviour of box and cylindrical columns and composite columns under biaxial bending conditions have been treated separately in Chapters 3 and 4 of *Axially Compressed Structures*.

7.2 DESIGN METHODS FOR BEAM COLUMNS

There are two basic design approaches used in the design of beam columns. The working stress (or allowable stress) design has been the principal approach used during the last century. The focus is placed on service load conditions. It is still the principal concept used in many specifications such as the AISC (American Institute of Steel Construction) Specifications. The other design philosophy may generally be referred to as limit state design. It is the basis of the Load and Resistance Factor Design (LRFD). Limit states may refer either to collapse or excessive elastic and inelastic deformations, or to the limit of serviceability such as cracking. Part 2 of the AISC Specifications for plastic design is such an example. It is a special case of limit state design since all limit states except the ultimate strength (or collapse load) have been precluded. In order to determine the ultimate strength of a beam column, inelastic properties must be taken into consideration.

For the allowable stress design procedure of beam columns, the AISC Specification can be considered as a typical example for discussion. The criteria can be converted directly from the ultimate strength interaction equations by discounting the factor of safety.

7.2.1 Plastic Design

(a) Yielding Interaction Criterion

For short columns or at support locations the strength interaction equation of the AISC Specifications is based on the consideration of yielding:

$$\frac{P}{P_y} + \frac{M}{1{\cdot}18M_p} \leq 1{\cdot}0 \qquad M \leq M_p \tag{7.1}$$

where M_p = full plastic moment = ZF_y, where Z = plastic section modulus and F_y = yield stress of the steel (kg in^{-2});
P_y = axial yield load;
P = maximum applied axial load = (load factor × actual load);
M = maximum applied moment.

In the presence of a compressive force P, the reduced plastic moment, M_{pc}, for a rectangular cross-section can be expressed as

$$\frac{M_{pc}}{M_p} = 1 - \left(\frac{P}{P_y}\right)^2 \tag{7.2}$$

Although eqn (7.2) is derived for a rectangular cross-section, it is also valid for all shapes including the W, S and M shapes. Equation (7.2) can be conservatively approximated by two linear curves:

$$\frac{M_{pc}}{M_p} = 1{\cdot}18\left(1 - \frac{P}{P_y}\right) \qquad \text{when} \qquad \frac{P}{P_y} > 0{\cdot}15 \tag{7.3a}$$

$$\frac{M_{pc}}{M_p} = 1 \qquad \text{when} \qquad \frac{P}{P_y} \leq 0{\cdot}15 \tag{7.3b}$$

Or upon rearranging,

$$\frac{P}{P_y} + \frac{M}{1{\cdot}18M_p} \leq 1{\cdot}0 \qquad \text{for} \qquad \frac{P}{P_y} > 0{\cdot}15 \tag{7.4a}$$

$$\frac{M}{M_p} \leq 1{\cdot}0 \qquad \text{for} \qquad \frac{P}{P_y} \leq 0{\cdot}15 \tag{7.4b}$$

where M_{pc} is replaced by the maximum applied moment M.

(b) Stability Interaction Criterion

For intermediate or slender columns the AISC ultimate strength interaction equation is given by

$$\frac{P}{P_{cr}} + \frac{C_m M}{\left(1 - \frac{P}{P_e}\right) M_m} \leq 1{\cdot}0 \tag{7.5}$$

where P_{cr} = the buckling strength of an axially loaded column;
M_m = the maximum bending moment that can be resisted by the member in the absence of axial load;
P_e = Euler buckling load of a column about the axis of bending.

C_m is a coefficient whose values are: (1) 0·85 for compression members in frames subjected to joint translation (sidesway); (2) for restrained compression members in frames braced against joint translation and not subjected to transverse loading between their supports in the plane of bending, $C_m = 0{\cdot}6 - 0{\cdot}4(M_1/M_2) \geq 0{\cdot}4$, where M_1/M_2 is the ratio of the smaller to larger moments; and (3) when the member is subjected to transverse loading, C_m can be determined either by a rational analysis or by the following approximate method:

$C_m = 0{\cdot}85$ for members without joint translation.
$C_m = 1{\cdot}00$ for members with joint translation.

For columns braced in the weak direction:

$$M_m = M_p \tag{7.5a}$$

For columns unbraced in the weak direction:

$$M_m = \left(1{\cdot}07 - \frac{\left(\frac{L}{r_y}\right)\sqrt{F_y}}{3160}\right) M_p \leq M_p \tag{7.5b}$$

where r_y = radius of gyration in the weak direction (inches). A member or a segment is defined as unbraced if the slenderness ratio of the laterally unsupported segment exceeds

$$\frac{L}{r_y} > \frac{1375}{F_y} + 25 \qquad \text{when} \qquad +1{\cdot}0 > \frac{M}{M_p} > -0{\cdot}5 \tag{7.6a}$$

or

$$\frac{L}{r_y} > \frac{1375}{F_y} \qquad \text{when} \qquad -0{\cdot}5 > \frac{M}{M_p} > -1{\cdot}0 \tag{7.6b}$$

where L = laterally unsupported length of the segment;
M = lesser of the moments at the ends of the unbraced segment;
M/M_p = end moment ratio (positive when the segment is bent in reverse curvature).

In eqn (7.5) P_{cr} is determined from the basic CRC (now SSRC) curve:

$$P_{cr} = P_e = \frac{\pi^2 EA}{\left(\frac{KL}{r}\right)^2} \qquad \text{when} \qquad \frac{KL}{r} > C_c \tag{7.7a}$$

which is the elastic buckling strength, where E = Young's modulus and A = cross-section area of the column, and

$$P_{cr} = P_y\left[1 - 0{\cdot}5\left(\frac{KL/r}{C_c}\right)^2\right] \qquad \text{when} \qquad \frac{KL}{r} < C_c \tag{7.7b}$$

for inelastic buckling, where KL/r is the effective slenderness ratio of the column, and

$$C_c = \sqrt{\left(\frac{2\pi^2 E}{F_y}\right)} \tag{7.7c}$$

is the slenderness ratio defining the elastic limit. The term $C_m/(1 - P/P_e)$ in eqn (7.5) is the amplification factor to account for the P–δ effect. For an elastic, simply supported beam column subjected to equal end moments M_0, M_{max} can be quite accurately expressed as $M_{max} = [1/(1 - P/P_e)]M_0$. For other transverse bending conditions, the equivalent moment concept (or the C_m method) is adopted so that M_{max} is equal to $C_m M_0$. The analytical solution of C_m for unequal end moments was derived by Massonnet (1959) with $C_m = [0{\cdot}3 + 0{\cdot}4(M_1/M_2) + 0{\cdot}3(M_1/M_2)^2]^{1/2}$. The current simplified design values for C_m in the AISC Specifications is C_m =

$0{\cdot}6 + 0{\cdot}4(M_1/M_2) \geq 0{\cdot}4$. This straight line falls near the upper limit for C_m at any given bending moment ratio M_1/M_2. The limiting value of 0·4 for C_m was established based on lateral–torsional buckling considerations.

When a beam column is subjected to transverse load rather than end moments, C_m may be approximated by $C_m = 1 + \psi(P/P_e)$, in which ψ is a function of end support conditions and depends on the loading type. For a simply supported prismatic member, ψ is equal to $(\pi^2\delta_0 EI/M_0L^2) - 1$, where δ_0 and M_0 represent the maximum deflection and moment between end supports due to transverse loading. Values of ψ for different loading conditions and end restraints are given in the commentary of the AISC Specifications. For unbraced frames, to account for the P–Δ effect, a value of 0·85 for C_m is used in the interaction equations.

The current plastic design method of the AISC Specifications can be applied to members in braced and unbraced multi-storey frames. However, a rational analysis which includes the effect of frame instability and column axial deformation is required in the determination of the ultimate strength of unbraced multi-storey frames.

Although many numerical solutions have been reported for simply supported beam columns under biaxial loading conditions, the labour involved in these calculations limits their practical use. Because the ultimate strength behaviour of biaxially loaded beam columns is not well understood, the current AISC Specifications do not provide a plastic design procedure for beam columns subjected to biaxial bending.

7.2.2 Allowable Stress Design

(a) Yielding Interaction Criterion

At support locations in braced frames and for members with very low slenderness ratios, yielding instead of instability governs the design. For such situations eqn (7.1) may be used with the introduction of a factor of safety, FS:

$$\frac{P_a/A}{P_y/(A\,.\,FS)} + \frac{M_a/S}{1{\cdot}18M_p/(S\,.\,FS)} = 1{\cdot}0 \qquad (7.8a)$$

or

$$\frac{f_a}{0{\cdot}60F_y} + \frac{F_b}{1{\cdot}18M_p/(S\,.\,FS)} = 1{\cdot}0 \qquad (7.8b)$$

where P_a and M_a are the service loadings for the beam columns.

The AISC Specifications use a conservative expression based on eqn (7.8b) by reducing the coefficient 1·18 to 1·0 and by substituting $M_p/(S\,.\,FS)$ by F_b. Therefore

$$\frac{f_a}{0{\cdot}6F_y} + \frac{f_b}{F_b} \leq 1{\cdot}0 \qquad (7{\cdot}9a)$$

For allowable stress design the biaxial loading cases can be checked by

$$\frac{f_a}{0{\cdot}6F_y} + \frac{f_{bx}}{F_{bx}} + \frac{f_{by}}{F_{by}} \leq 1{\cdot}0 \qquad (7.9b)$$

where $f_a = P_a/A$ = nominal axial compression stress at service load;
f_b, f_{bx}, f_{by} = flexural stress at service load based on primary bending moment about the x and y axes respectively;
F_b, F_{bx}, F_{by} = allowable flexural stresses for bending about the x and y axes respectively.

(b) Stability Interaction Criterion
The ultimate strength interaction equation including lateral–torsional buckling is expressed by eqn (7.5). It may be written as:

$$\frac{P_a/A}{P_{cr}/(A\,.\,FS)} + \frac{C_m}{\left[1 - \dfrac{P_a/A}{P_e/(A\,.\,FS)}\right]} \frac{M_a/S}{M_m/(S\,.\,FS)} \leq 1{\cdot}0 \qquad (7.10a)$$

or

$$\frac{f_a}{F_a} + \frac{C_m}{\left(1 - \dfrac{f_a}{F'_e}\right)} \frac{f_b}{F_b} \leq 1{\cdot}0 \qquad (7.10b)$$

where F'_e is the Euler buckling stress for a prismatic member about its bending axis divided by FS, and F_a is the allowable compressive stress considering the member as an axially loaded column only. For biaxial bending and compression:

$$\frac{f_a}{F_b} + \frac{C_{mx}}{\left(1 - \dfrac{f_a}{F'_{ex}}\right)} \frac{f_{bx}}{F_{bx}} + \frac{C_{my}}{\left(1 - \dfrac{f_a}{F'_{ey}}\right)} \frac{f_{by}}{F_{by}} \leq 1{\cdot}0 \qquad (7.10c)$$

(c) Simplified Interaction Criterion for Small Axial Load
When the applied axial load is sufficiently small so that f_a/F_a does not exceed 0·15, the AISC Specification permits the use of the following

formula instead of eqns (7.9b) and (7.10b):

$$\frac{f_a}{F_a} + \frac{f_{bx}}{F_{bx}} + \frac{f_{by}}{F_{by}} \leq 1{\cdot}0 \tag{7.10d}$$

7.2.3 Other Design Formulae

There are a large number of design equations for beam columns in the USA and in other countries. They more or less use the interaction equation format of the AISC Specifications, although many details are different. Different values of the factor of safety and the use of multiple column curves for the allowable axial stress are two such examples.

7.3 DISCUSSION OF THE BEAM COLUMN INTERACTION EQUATIONS

The interaction equations of the AISC type for beam column design are conceptually simple, convenient to use and have been applied in a wide range of situations. Although the interaction equations were derived originally for members in a braced frame subjected to in-plane bending only, they have been extended to cover most other failure modes. It is because of the latter that this design practice contains certain undesirable features that have been discussed by many researchers. They are described in the following.

7.3.1 Factor of Safety

In the interaction equations for allowable stress design, the factor of safety for F_a varies with the value of the slenderness ratio from 1/1·67 to 1/1·92. The factor of safety for F_b is usually 1/1·67 of the first yield moment. However, when the failure mode is lateral–torsional buckling, the factor of safety for F_b is difficult to define. Therefore, the factor of safety of beam columns cannot be examined and explained on a rational basis. This is worse for the case of biaxially loaded members. By comparing published results with their own, Pillai and Ellis (1974) have reported that the interaction equation, eqn (7.10c), for biaxially loaded members is conservative over a wide range of conditions.

When lateral–torsional buckling governs the design of beam columns, Galambos (1963) has shown that eqn (7.5b) is a realistic expression for plastic design. For allowable stress design based on the lateral–torsional buckling strength the current AISC practice, by providing two equations and permitting the use of only a fraction of the total strength, represents an underestimate of the computational capacity of structural engineers.

7.3.2 Nature of the Interaction Equations

(a) Linear Interaction Equations

The beam column interaction equations either for allowable stress design or for ultimate strength design have a linear form in which a direct relationship exists between the axial load and the amplified bending moment. However, a recent study by Cheong-Siat-Moy and Downs (1979) shows that the ultimate strength interaction relationship between P and M is clearly non-linear.

(b) P–Δ Effect

The effect of frame instability on the behaviour of beam columns in unbraced frames can be evaluated either accurately by a complete second-order elastic analysis, or approximately by using the methods proposed by LeMessurier (1977) and Cheong-Siat-Moy and Downs (1979). These efforts have directed attention to the two confusing elements in the interaction equation: $C_m = 0{\cdot}85$ and the estimation of the various effective length factors for an unbraced frame.

K-Factor

In the interaction equations the K-factor appears in the first term for axially loaded columns and again in the amplification factor. A theoretical study by Yura and Galambos (1965) has shown that the introduction of K into the amplification factor for columns in a single-storey frame is a reasonable procedure to account for the P–Δ effect. However, rigid floor diaphragm action has an effect on the K values of the columns by providing the same restraining effect on all columns on the same floor. One storey of a structure may buckle as a unit at the failure load. It is therefore more reasonable to consider K for the buckling load of that storey. In addition, there are some other factors which can influence the K value such as the flexible connection and the partial base fixity of the frame.

(d) Lateral–Torsional End Restraints

The AISC interaction equations were established originally based on the in-plane strength of beam columns. With some modifications for the value of F_b and M_m in eqns (7.10b) and (7.5b), they are adjusted to accommodate the out-of-plane failure mode. The in-plane instability and the out-of-plane buckling of a beam column are described by different mathematical models with different sets of parameters. Other than for reason of simplicity, there is a lack of a rational basis for using the same interaction equations for the design against both the in-plane and the out-of-plane failure modes.

For in-plane failure, the end restraining effects can be considered in F_a and in F_e' because they are axial buckling stresses. For out-of-plane failure there is a problem in F_b, because the lateral buckling stress is derived from simply supported conditions. Since most laterally unsupported segments are portions of girders, they should be viewed as laterally continuous beams. This conservative measure in determining F_b is not only unrealistic, but also contributes to the confusion in beam column design. (Take into consideration the end restraining effect in F_a but not in F_b.) This question has been examined by several researchers in recent years.

7.4 ALTERNATIVE APPROACHES TO BEAM COLUMN DESIGN

7.4.1 Modified Effective Length Approach

In typical engineering practice the effect of instability is not handled in the structural analysis. Rather, it is compensated for, approximately, by proportioning the individual structural members according to instability criteria with appropriate end conditions including the possibility of sidesway. Furthermore, the coefficient C_m is set to a value of 0·85. To evaluate the P–Δ effect in the analysis of structures is a straightforward procedure. However, how to include the frame instability effect in the interaction design equations in a rational and simple fashion remains to be satisfactorily answered.

Based on the assumption that the P–Δ effect is more predominant than the P–δ effect in sway frames, LeMessurier (1977) proposed the following stability interaction criterion for beam column design:

$$\frac{P}{P_{cr}} + \frac{1}{1 - \dfrac{\sum P}{\sum P_L - \sum (C_L P)}} \frac{M}{M_m} \leq 1{\cdot}0 \tag{7.11a}$$

or

$$\frac{P}{P_{cr}} + A_F \frac{M}{M_m} \leq 1{\cdot}0 \tag{7.11b}$$

where

$$A_F = \frac{1}{1 - \dfrac{\sum P}{\sum P_L - \sum (C_L P)}} = \text{the amplification factor} \qquad (7.11c)$$

and $\sum P$ = total gravity load on the storey;
C_L = factor accounting for the reduction in column rotation stiffness due to the presence of axial load P;
P_L = a measure of the stiffness of the structure subjected to lateral load.

Furthermore, since all columns in a storey can in most cases be assumed to buckle simultaneously, he proposed that the effective length factor K should be calculated based on such a philosophy. The modified K-factor for the ith column of the storey recommended is given by

$$K_i^2 = \frac{\pi^2 I_i}{P_i} \left[\frac{\sum P + \sum (C_L P)}{\sum (\alpha I)} \right] \qquad (7.12)$$

where I_i = moment of inertia of the ith column;
P_i = axial load on the ith column;
α = factor reflecting the effect of the various restraining conditions at the ends of the individual columns.

For example, when the column has a point of inflection at middle height,

$$\alpha = \frac{12}{1 + G}$$

$$C_L \simeq \frac{0{\cdot}22}{(1 + G)^2}$$

$$P_L = \frac{Qh}{\Delta_F}$$

where $G = \sum (I_c/L_c)/\sum (I_b/L_b)$ = ratio of column stiffness to beam stiffness;
Q = lateral force on the floor;
Δ_F = lateral displacement of the floor;
h = half length of the column.

The K value determined by eqn (7.12) is used to calculate P_{cr} in eqn (7.11). However, similar to the AISC interaction equation, eqn (7.11) also has a linear relationship between P and the amplified moment. The two main differences between the LeMessurier interaction equation and the AISC interaction equation are (1) the method of determining the effective length factors, and (2) the method to account for the P–Δ effect. LeMessurier's approach considers both aspects simultaneously in one rational process.

7.4.2 Non-Linear Interaction Equations

(a) Uniaxial Bending

The above discussion is concerned with linear interaction equations for beam column design. Many experimental results have shown that the ultimate strength relationship between P and the amplified moment is non-linear. It was also observed from these results that the non-linearity increases as the P–Δ effect increases. Based on this information, a non-linear interaction equation has been proposed by Cheong-Siat-Moy and Downs (1979):

$$\frac{P}{P_{cr}} + \beta A_F \frac{M}{M_m} \leq 1{\cdot}0 \tag{7.13}$$

where $\beta = 0{\cdot}9 + 0{\cdot}1 A_F$;
A_F = as defined in eqn (7.11c);
$K = 1{\cdot}0$ in the calculation of critical buckling loads.

This equation is valid only for combined gravity and lateral loads. It is inappropriate for situations where only gravity loads exist. For this latter case a separate conservative formula is proposed which is similar to eqn (7.5):

$$\frac{P}{P_{cr}} + \frac{C_m}{1 - \dfrac{P}{P_e}} \frac{M_g}{M_m} \leq 1{\cdot}0 \tag{7.14}$$

where M_g = moment due to gravity load.

Since, in most cases, moments in structures are due to both gravity and lateral loads, a modified form of eqn (7.13) is also proposed by Cheong-Siat-Moy and Downs:

$$\frac{P}{P_{cr}} + \frac{\beta C_m}{1 - \dfrac{P}{P_e}} \left(\frac{A_F M_0 + M_g}{M_m} \right) \leq 1{\cdot}0 \tag{7.15}$$

where M_0 = moment induced by lateral loads.

The corresponding interaction equations for allowable stress design are given by:

For combined loads:

$$\frac{f_a}{F_a} + \beta \frac{C_m}{1 - \dfrac{P}{P_e}} \left(\frac{A_F f_b + f_{bg}}{F_b} \right) \leq 1{\cdot}0 \tag{7.16a}$$

with $K = 1{\cdot}0$.

For gravity loads:

$$\frac{f_a}{F_a} + \frac{C_m}{1 - \dfrac{P}{P_e}} \frac{f_{bg}}{F_b} \leq 1{\cdot}0 \tag{7.16b}$$

with K determined from $K^2 = P_e(1 - [1/A_F])/P$ or more rigorous analysis.

In eqns (7.15) and (7.16a) the amplification factor A_F should be calculated based on the factored load. For standard frames, A_F may be calculated from eqn (7.11c) with satisfactory results. For irregular structures a second-order elastic analysis is recommended to determine the amplified moment.

(*b*) *Biaxial Bending*

Strictly speaking, most beam columns are subjected to biaxial bending in addition to the axial load. Equations (1) and (5) of the AISC Specifications have been extended to cover the design of biaxially loaded columns (SSRC, 1976) as described in the following.

For short columns where the P–δ effect is negligible:

$$\frac{P}{P_y} + \frac{M_x}{1{\cdot}18 M_{px}} + \frac{M_y}{1{\cdot}18 M_{py}} \leq 1{\cdot}0 \tag{7.17a}$$

For long columns where the P–δ effect is significant:

$$\frac{P}{P_{cr}} + \frac{C_{mx}}{1 - \dfrac{P}{P_{ex}}} \frac{M_x}{M_{mx}} + \frac{C_{my}}{1 - \dfrac{P}{P_{ey}}} \frac{M_y}{M_{my}} \leq 1{\cdot}0 \tag{7.17b}$$

An alternative expression for these equations can be obtained by using eqns (7.2) and (7.3) and by some mathematical manipulations as well as the introduction of a minor modification for M_{pcy}:

$$\frac{M_x}{M_{pcx}} + \frac{M_y}{M_{pcy}} \leq 1{\cdot}0 \tag{7.18a}$$

and

$$\frac{C_{mx}M_x}{M_{ucx}} + \frac{C_{my}M_y}{M_{ucy}} \leq 1{\cdot}0 \tag{7.18b}$$

where

$$M_{pcx} = 1{\cdot}18M_{px}\left(1 - \frac{P}{P_y}\right) \leq M_{px}$$

$$M_{pcy} = 1{\cdot}19M_{py}\left[1 - \left(\frac{P}{P_y}\right)^2\right] \leq M_{py}$$

$$M_{ucx} = M_{mx}\left[1 - \frac{P}{P_{cr}}\right]\left[1 - \frac{P}{P_{ex}}\right]$$

$$M_{ucy} = M_{py}\left[1 - \frac{P}{P_{cr}}\right]\left[1 - \frac{P}{P_{ey}}\right]$$

$$M_{mx} = M_{px}\left[1{\cdot}07 - \frac{(L/r_y)F_y}{3160}\right] \leq M_{px}$$

It should be noted that the above are still linear interaction equations. Research has shown that the interactive nature of moments about the orthogonal axes is non-linear and that they resemble closely the quadrant of a circle. For a better representation of the interactions between moments, a non-linear expression of the interaction equations has been proposed by Tebedge and Chen (1974):

$$\left(\frac{M_x}{M_{pcx}}\right)^a + \left(\frac{M_y}{M_{pcy}}\right)^a \leq 1{\cdot}0 \tag{7.19a}$$

$$\left(\frac{C_{mx}M_x}{M_{ucx}}\right)^b + \left(\frac{C_{my}M_y}{M_{ucy}}\right)^b \leq 1{\cdot}0 \tag{7.19b}$$

The exponents a and b in eqn (7.19) can be obtained by fitting the actual strength curve of biaxially loaded beam columns with different cross-sections. For wide-flange shapes having flange width B and web depth D, the exponents may be expressed as follows:

At braced location:

$$a = 1\cdot6 - \frac{(P/P_y)}{2\ln(P/P_y)} \tag{7.20a}$$

Between bracing points:

$$b = 0\cdot4 + \frac{P}{P_y} + \frac{B}{D} \geq 1\cdot0 \qquad \text{for} \qquad \frac{B}{D} \geq 0\cdot3 \tag{7.20b}$$

$$b = 1\cdot0 \qquad \text{for} \qquad \frac{B}{D} < 0\cdot3 \tag{7.20c}$$

For a square box column:

$$a = 1\cdot7 - \frac{(P/P_y)}{\ln(P/P_y)} \tag{7.21a}$$

$$b = 1\cdot3 + \frac{1000(P/P_y)}{(L/r)^2} \geq 1\cdot4 \tag{7.21b}$$

The advantage of the non-linear design formulae proposed by Chen and his associates is that they have universal applicability for different shapes of cross-section. When the exponents of a particular cross-section are not determined, they can be assumed to be unity so that eqn (7.18) results. For uniaxial bending conditions, eqn (7.19) is also applicable if bending in one of the two directions is set to zero. The only restriction of these non-linear interaction equations is that they are applicable to non-sway beam columns. Although a modified form has also been proposed (Chen and Cheong-Siat-Moy, 1980) to account for the P–Δ effect, this latter formula is said to be less reliable and further validations are desirable.

7.4.3 Separation of In-Plane and Out-of-Plane Failure Modes

The theoretical basis for the current AISC interaction equation is the in-plane beam column behaviour. This basis was then modified based on experimental evidence and other considerations so that it is used to accommodate the design against lateral–torsional buckling. This approach has been used widely and satisfactorily for many years by structural engineers. In an effort to establish a more rational basis for allowable stress design, Hsu and Lee (1981) have carried out an analytical study and suggested that beam columns may be designed with two separate governing interaction equations, one for the in-plane and the other for the out-of-plane failure modes. They concluded that the current AISC equation is most suitable for the in-plane behaviour, or for beam columns laterally braced. In that case P_{cr} and P_e in the amplification factor are the same (strong axis buckling).

For lateral–torsional buckling, one may use the basic linear interaction equation

$$\frac{P}{P_{cr}} + \xi \frac{M}{M_m} \leq 1{\cdot}0 \tag{7.22a}$$

or

$$\frac{f_a}{F_a} + \xi \frac{f_b}{F_b} \leq 1{\cdot}0 \tag{7.22b}$$

where P_{cr} = the weak axis buckling load (out-of-plane buckling);

$$\xi = 1 - \frac{1}{2}\frac{P}{P_{cr}} \quad \text{or} \quad 1 - \frac{1}{2}\frac{f_a}{F_a} \qquad \text{for} \quad \frac{L}{r_y} \geq 100$$

$$= 1 \qquad \text{for} \quad \frac{L}{r_y} < 100$$

$$F_b = \frac{2}{3}\left[1{\cdot}0 - \frac{F_y}{6{\cdot}435\sqrt{(F_{b1}^2 + F_{b2}^2)}}\right] F_y \qquad \text{for} \quad 0{\cdot}42F_y \leq F_b \leq 0{\cdot}6F_y$$

which is the standard transition curve from $0{\cdot}7F_y$ to $1{\cdot}11F_y$ (with assumed maximum compressive residual stress equal to $0{\cdot}3F_y$);

$$F_b = \sqrt{(F_{b1}^2 + F_{b2}^2)} \qquad \text{for} \quad F_b < 0{\cdot}42F_y$$

$$F_{b1} = \frac{170 \times 10^3 C_b}{\left(\dfrac{K_w L}{r_t}\right)^2}$$

$$F_{b2} = \frac{12 \times 10^3 C_b}{\left(\dfrac{K_s L d}{A_f}\right)}$$

$$C_b = 1{\cdot}75 - 1{\cdot}05\frac{M_1}{M_2} + 0{\cdot}3\left(\frac{M_1}{M_2}\right)^2 = 1{\cdot}75 \text{ for concentrated load}$$

K_w = effective length factor of laterally continuous beams of thin and deep type of cross-section where warping torsional resistance is important;
K_s = effective length factor of laterally continuous beams of thick and shallow type of cross-section where the St Venant torsional resistance is important;
A_f = area of flange;
d = depth of section.

The curves for the effective length factors K_w, K_s are given in Fig. 7.2(a) and (b). The details for the derivation of these factors are given by Hsu and Lee

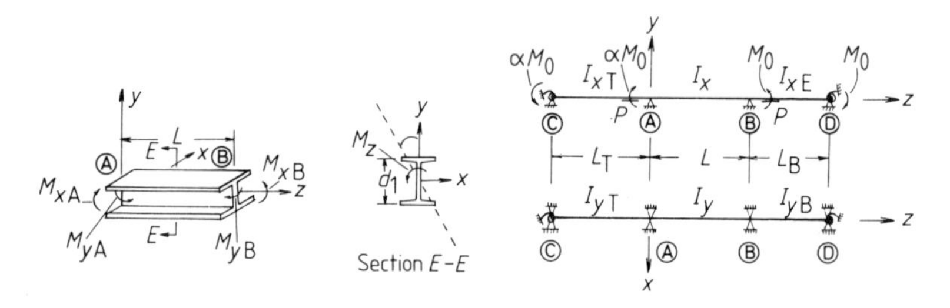

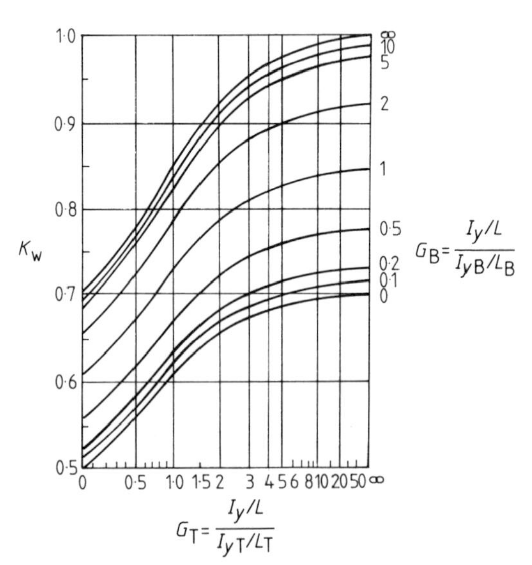

FIG. 7.2(a). Effective length factors, K_w, for the 'warping' term of laterally unsupported beams.

(1981). They provide welcome information to designers for the determination of allowable bending stress for laterally continuous beams. Such information is lacking in the current AISC design Specifications. The determination of ξ in eqn (7.24) is based on curve fittings of analytical results obtained for the lateral buckling strength (both elastic and inelastic) for various beam column cross-sections with different end moment ratios, slenderness ratios and P/P_{cr} ratios.

For the purpose of comparison of eqn (7.22) with eqn (7.5), an elastic, simply supported beam column with I-section may be examined, and the

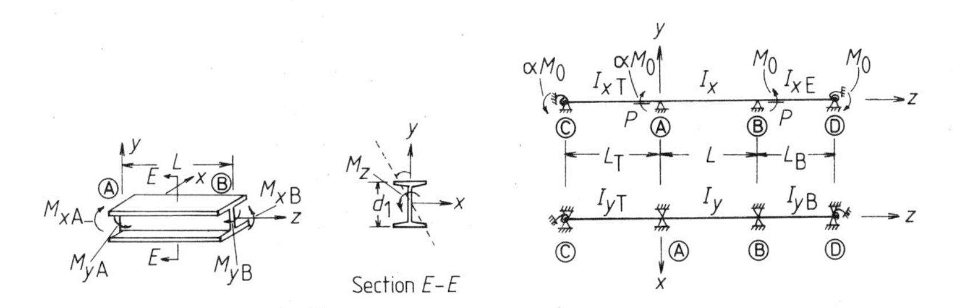

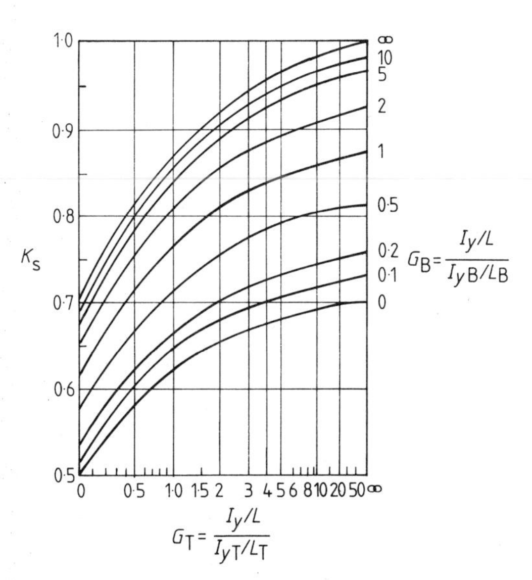

FIG. 7.2(b). Effective length factors, K_s, for the 'St Venant' term of laterally unsupported beams.

predictions of these two equations together with that of the exact solution can be compared.

To do this, eqns (7.5), (7.22a) and the exact solution for an elastic, simply supported beam column subjected to equal end moment can be written as those expressed by eqns (7.23a), (7.23b) and (7.23c):

$$\frac{P}{(P_{cr})_y} + \frac{M}{\left(1 - \dfrac{P}{P_{ex}}\right) M_m} \leq 1{\cdot}0 \tag{7.23a}$$

$$\frac{P}{(P_{cr})_y} + \left(1 - \frac{1}{2}\frac{P}{P_{ey}}\right)\frac{M}{M_m} \leq 1{\cdot}0 \tag{7.23b}$$

$$\sqrt{\left(\left[1 - \frac{P}{P_{ey}}\right]\left[1 - \frac{P}{P_{ez}}\right]\right)} - \frac{M}{M_m} \leq 1{\cdot}0 \tag{7.23c}$$

where $P_{ex} = \dfrac{\pi^2 EI_x}{L^2}$

$$P_{ey} = \frac{\pi^2 EI_y}{L^2}$$

$$P_{ez} = \frac{A}{I_x + I_y}\left(GK_T + \frac{\pi^2}{L^2} EI_w\right)$$

Assuming that for most wide flange shapes the ratios of $P_{ex}:P_{ey}:P_{ez}$ fall between 20:1:2 and 3:1:2, eqns (7.23a), (7.23b) and (7.23c) may be quite simply compared. These are shown in Fig. 7.3. This simple comparison suggests that eqn (7.22) is a closer approximation of the exact solution and that eqn (7.5) is more conservative. Obviously, eqn (7.5) will result in yet more conservative designs for laterally continuous beam columns.

7.5 SUMMARY

In this brief discussion on the design of beam columns, the AISC interaction equations are used as the basis because they are widely used, conceptually simple and easy to use. Certain other approaches are also briefly described to illustrate the scope of the current status and research interest in beam column design.

There are a large number of current research activities, publications and design guide formulations not mentioned in this discussion; for example,

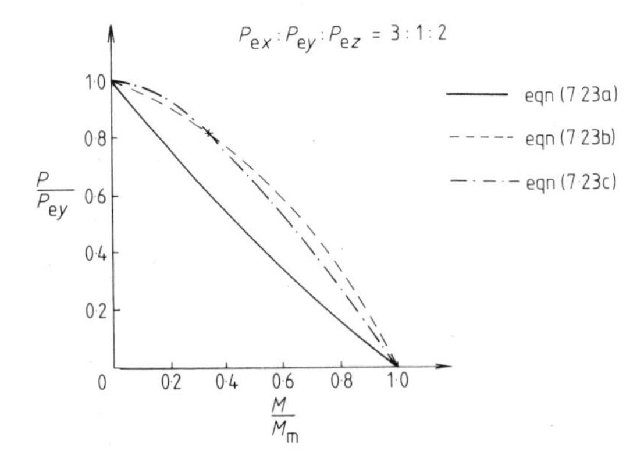

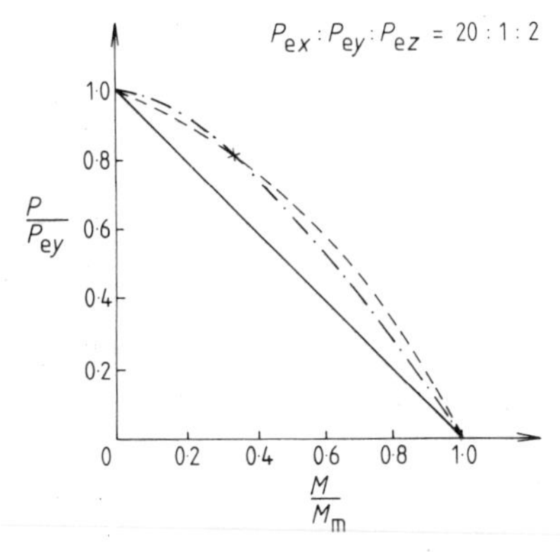

FIG. 7.3. Comparison of eqns (7.23a), (7.23b) and (7.23c) for elastic, simply supported beam columns.

the use of multiple column curves for the allowable axial stress, the use of tapered beam columns and hybrid beam columns, substructural analysis, etc. Many of these studies have been incorporated into design specifications in various countries of the world.

Because of the rapid advances made in non-linear analysis in computational methods, and in computer capacity, it is reasonable to expect that future development in the design of beam columns may undergo some qualitative changes rather than the continuation of refining the current practice in a quantitative sense. If this were true, then it appears that continued effort devoted to such issues as establishing the most desirable set of multiple column curves for the allowable axial stress is less important and unproductive. On the other hand, problems concerning the ultimate strength and instability behaviour of biaxially loaded, restrained beam columns subjected to various combinations of loading conditions are still not clearly understood. This is an important area because it provides a sound basis for potential future computer-aided design practice which conceivably will deal with three-dimensional analysis and design of structures directly, including the P–Δ and other instability effects.

Another challenging research area of beam column design is the investigation of the effect of repeated loading as well as non-proportional loading on the behaviour of beam columns, particularly for large inelastic deformations. Such research results can greatly facilitate the design of structures against strong earthquake ground motions where the design philosophy requires the understanding of the maximum deformability of structures subjected to non-proportionally applied loading sequences.

REFERENCES

CHEN, W. F. and ATSUTA, T. (1976) *Theory of Beam Columns*, Vol. 1: *In-Plane Behavior and Design*, McGraw-Hill, New York.

CHEN, W. F. and ATSUTA, T. (1977) *Theory of Beam Columns*, Vol. 2: *Space Behavior and Design*, McGraw-Hill, New York.

CHEN, W. F. and CHEONG-SIAT-MOY, F. (1980) Limit states design of beam-columns. *SM Archives*, **5**(1), 29–73.

CHEONG-SIAT-MOY, F. and DOWNS, T. (1979) New interaction equation for steel column design. Report No. C1, Dept of Civil and Mineral Engineering, University of Minnesota (Mar.).

GALAMBOS, T. V. (1963) Inelastic lateral buckling of beams. *J. Struct. Div., Proc. ASCE*, **89**(ST5), 217–42.

HSU, T. L. and LEE, G. C. (1981) Design of beam-columns with lateral-torsional end restraints. *WRC Bulletin*, No. 272, 1–14.

LEMESSURIER, W. J. (1977) A practical method of second order analysis, Part 2: Rigid frame. *Engineering Journal, AISC*, **14**(2), 49–67.

MASSONNET, C. (1959) Stability considerations in the design of steel columns. *J. Struct. Div., Proc. ASCE*, **85**(ST7), 75–111.

PILLAI, U. S. and ELLIS, J. S. (1974) Beam-columns of hollow structural sections. *Can. J. Civil Engng.*, **1**, 194–8.

SSRC (1976) Structural Stability Research Council, *Guide to Stability Design Criteria for Metal Structures*, 3rd Edn, ed. B. J. Johnston, John Wiley, New York.

TEBEDGE, N. and CHEN, W. F. (1974) Design criteria for H-columns under biaxial bending. *J. Struct. Div., Proc. ASCE*, **100**(ST3), 579–98.

YURA, J. A. and GALAMBOS, T. V. (1965) Strength of single-storey steel frames. *J. Struct. Div., Proc. ASCE*, **91**(ST5), 81–101.

Chapter 8

TRENDS IN SAFETY FACTOR OPTIMISATION

N. C. Lind

Department of Civil Engineering, University of Waterloo, Ontario, Canada

and

M. K. Ravindra

Structural Mechanics Associates, Newport Beach, California, USA

SUMMARY

Selection of safety factors in a design code is an important decision made by the code writers. It has significant effects on the cost and reliability of future designs produced using the code. Therefore, it is essential to make an objective evaluation of these effects such that the selection of safety factors can be made optimal.

An overall framework for deriving optimal safety factors, called 'code optimisation', was first proposed by Ravindra and Lind (1973). A code is idealised as a set of code parameters (e.g. safety factors, nominal material strengths and basic wind pressures). A specific set of code parameter values describes a code in a mathematical sense. An expected cost of such a code is defined as the sum of initial costs of all structures designed using the code and the mathematical expectation of the costs of failure of all these structures. Code optimisation is formulated as a minimisation of this scalar point function (i.e. total expected cost) and is shown to be reducible to a problem in non-linear programming. Application of this theory on a limited scale has resulted in the development of the National Building Code of Canada, the

Ontario Highway Bridge Code and the American National Standards Institute's Loading Standards.

The central theme of code optimisation is that the selection of parameter values in a code should take into account all the structures to be designed in future using the code. Another important concept is that probabilistic design can be accomplished in a deterministic format. These concepts are utilised at present in selecting optimal load and resistance factors in a specified set of load combinations. The objective is to achieve reliabilities for different structures as close to the target values as optimally possible. Examples of this application are described by Ravindra et al. (*1974*), *Nowak and Lind* (*1979*), *Ellingwood* et al. (*1980*) *and Schwartz* et al. (*1981*).

NOTATION

c_f, $c_i(\omega, \zeta, \mathbf{p})$	Cost of failure and initial cost of structure
$C_I(\mathbf{p})$, $C_F(\mathbf{p})$, $C_T(\mathbf{p})$	Initial, expected failure and total cost of code, respectively
C_0	Maintenance and demolition cost
D	Dead load
E	Earthquake load
Ec_f	Total expected cost of failure of a structure
$f(\;)$	Frequency density of occurrence of a continuous random variable
L	Live load
M_p	Nominal plastic moment capacity of cross-section
$\mathbf{M}$	Member proportion set
$\mathbf{p}$	Code parameter set
p_F	Probability of failure
$\mathbf{P}$	Code parameter space
Q, Q_n, Q_{apt}	Load, nominal load and 'arbitrary point in time' load, respectively
R	Resistance of structural element; R_n is nominal value of R
S	Snow load
$\mathbf{S}_i$	Space of structural schemes
T	Thermal load
W	Wind load
X_i	Random variable; mean m_{X_i}, standard deviation σ_{X_i} and coefficient of variation V_{X_i}

Y	Response
$\mathbf{Z}$	Designer's choice space
$Z(\phi, \gamma)$	Objective function
β	Reliability index; β_0 = target value of β
γ	Load factors, γ_D, γ_L, etc.
γ_I	Importance factor
ζ	Set of designer's choices
θ	Set of real physical variables
Θ	Space of real physical variables
Λ	Member proportion space
ϕ	Resistance factor
ψ	Load combination factor
ρ	Coefficient of correlation
Ω	Data space
ω	Data set

8.1 INTRODUCTION

8.1.1 Safety Factors: Need

Safety factors are utilised in structural design codes to ensure that the nominal resistance of a structural element (i.e. beam, column, floor slab, sub-assembly, etc.—or an entire structure) exceeds the calculated load effect (e.g. moment, stress, deflection, overturning moment, etc.) acting on it. Safety factors are intended to account for variabilities in material strengths and imposed loads, and uncertainties introduced by design assumptions and construction practices. The values of the safety factors are selected such that the resulting designs have an acceptably low probability of failure.

A single global safety factor was specified for each of a selected set of structural elements in some of the early design codes. Many such codes are still in use. For example, a safety factor of 1·67 is used in the design of steel structures (AISC, 1969). Designers have come to realise that a single safety factor cannot adequately represent the myriad design situations (i.e. different structural configurations, materials of construction, building occupancies, etc.) in which the element may occur. Also, a single factor may not be appropriate for different loads exhibiting markedly different variabilities. Therefore, the current trend is to specify a number of partial safety factors as exemplified in the ACI (1977) Building Code Requirements, the National Building Code of Canada (1980) and the Nordic Committee on Building Regulations (1978).

Safety factors also fulfil the practical need of maintaining design rules that are deterministic in form but based on a probabilistic rationale. As pointed out by Lind (1969), probabilistic design codes are more readily accepted by the profession if they have a deterministic appearance. Also, the designer may not have all the statistical data needed to perform a probabilistic design and may not be equipped to make the required judgemental decisions with regard to lack of data and lack of probabilistic models. Code writers, on the other hand, are charged with the responsibility to develop design rules making use of all available data, analytical models, and judgement based on the collective experience of the profession. As a result, they may specify the values of the partial safety factors to be used in specific load combinations for the design of structural elements. Such a set of design rules is known as a Level I code. Higher levels of codes have been discussed by Rackwitz (1976). However, in the foreseeable future, the use of partial safety factors in a limit states design code format will continue.

8.1.2 Safety Factors and Gross Errors

Control of the reliability of structures requires not only adequate safety factors, but also appropriate inspection in design and construction. Recent comprehensive studies of structural failures indicate that gross error is present in the majority of cases and may be the cause of more than 80 % of the failures (Matousek, 1977). Safety factors may contribute modestly to reduce failures due to minor human error. But, however generous, they are ineffective in preventing failure due to gross human error. Safety factors are effective in the control of failures due to uncertainties in loads, resistances, methods of analysis and normal fluctuations of workmanship. Inspection, material control and checking of design are effective in the control of gross error, but ineffective against load and strength uncertainties. It is sometimes argued that modern building codes, with their elaborate systems of safety factors, are thus concentrating on the control of only a small fraction of potential failures. Further, it is argued that probabilistic rationales neglect gross error, the most common cause of failure, and are therefore an inappropriate basis for structural regulations. These arguments are fallacious; a comparison of *absolute* failure frequencies is inappropriate. Each measure of control should be assessed in terms of the *marginal* returns on safety for an incremental investment. If indeed a small sum could be saved by lowering a safety factor, and then reinvested in the safety of a structure by way of improved inspection so as to effect a net increase in reliability, then that safety factor is too high. However, it is clear that the potential amount to be saved by such reductions of safety margins is limited, and at some point the number of failures prevented by additional inspection would become insufficient to offset the increase in failures of well-designed structures. Codes are primarily designed to control the reliability of structures without gross error; they have been highly successful in this task, as witnessed by observed failure rates, and their optimisation by probabilistic analysis is the rational way towards improvement.

8.1.3 Code Formats

A number of code formats have been proposed in recent years for probabilistic limit states design. They all conform to the general principle that, for each limit state, the following inequality should hold:

$$\text{Factored resistance} \geq \text{Sum of factored load effects} \qquad (8.1)$$

The limit states considered are ultimate limit states (e.g. plastic collapse, instability, rupture due to fatigue, and loss of equilibrium), conditional limit states (e.g. local failure due to accidents) and serviceability limit states (e.g. excessive deflection, local damage, and excessive vibration producing discomfort). For a given limit state, many limit state checking equations may have to be satisfied corresponding to the postulated load combinations. The National Building Code of Canada (1980) uses the following format for its limit states design criteria:

$$\phi R_n \geq \gamma_D D_n + \gamma_I \psi (\gamma_L L_n + \gamma_W W_n + \gamma_T T_n) \qquad (8.2)$$

in which ϕ is the resistance factor, R_n is the nominal resistance of the structural element, D_n, L_n, W_n, T_n are nominal dead, live, wind and imposed deformation loads; $\gamma_D, \gamma_L, \gamma_W, \gamma_T$ are load factors, γ_I is an importance factor reflecting use and occupancy of the building and ψ is a load combination factor designed to reflect the smaller probability that two or more loads will attain their design values simultaneously. The Comité Euro-International du Béton (1976) has proposed the format:

$$\phi R_n \geq \gamma_D D_n + \gamma_Q (Q_{ni} + \sum \psi_{0j} Q_{nj}) \qquad (8.3)$$

in which Q_{ni} is the principal variable load, and ψ_{0j}, Q_{nj} are frequently occurring values of the accompanying loads ($\psi_{0j} < 1{\cdot}0$). Different load combinations are generated by rotating the individual time-varying loads in eqn (8.3), each taking the position of the principal variable load Q_{ni} while the remaining loads are assigned values, Q_{nj}. The selected nominal resistance of the structural element should satisfy eqn (8.3) for all load combinations.

The Load and Resistance Factor Design (LRFD) format proposed for structural steel design in the USA (Galambos and Ravindra, 1978) is:

$$\phi R_n \geq \gamma_D D_n + \gamma_Q Q_n + \sum \gamma_{apt} Q_{apt} \tag{8.4}$$

in which Q_{apt} are the frequently occurring 'arbitrary point in time' loads and γ_{apt} are the load factors applied on Q_{apt}.

Most recently, the Canadian Standards Association (1981) has specified the following general format as a common basis for design and evaluation of civil engineering structures:

$$\phi_j R_j(\phi_y f'_y, \phi_c f'_{c}, \ldots, d' - \Delta, \ldots) \geq \sum \gamma_{ij} G_i + \sum \gamma_{ij} Q_i + A_j \tag{8.5}$$

in which f'_y, etc., are specified material properties, d' specified dimensions, while G, Q and A are dead, live and accidental loads, respectively, which in the various combinations are at so-called specified, frequent or sustained level values.

In the above equations, the resistance factors ϕ are intended to account for the possible unfavourable deviations of the strength of the materials and other properties from nominal values, for the uncertainties introduced by fabrication, and for the inaccuracies in the calculation model for structural resistance. The load factors γ are intended to account for unforeseen and unfavourable deviations in the actual loads from their nominal values, Q_n or Q_{apt}, and for variations in the load effects due to uncertainties in the structural analysis. For a detailed description of the sources of these variabilities, the reader is referred to Galambos and Ravindra (1978) and MacGregor (1976).

8.1.4 The Second Moment Rationale

Development of load and resistance factors is accomplished in recent design codes using first-order second moment reliability theory. The theory, first proposed in essence by Mayer (1926) and later independently by others, was developed independently in practically workable form by Cornell (1969). Lind (1969) demonstrated the derivation of load and resistance factors by second moment theory. The theory is based on the concept that the random variables of interest to structural safety can be modelled by their first two statistical moments, viz. mean and variance. The mean of a random variable indicates the central tendency of the variable

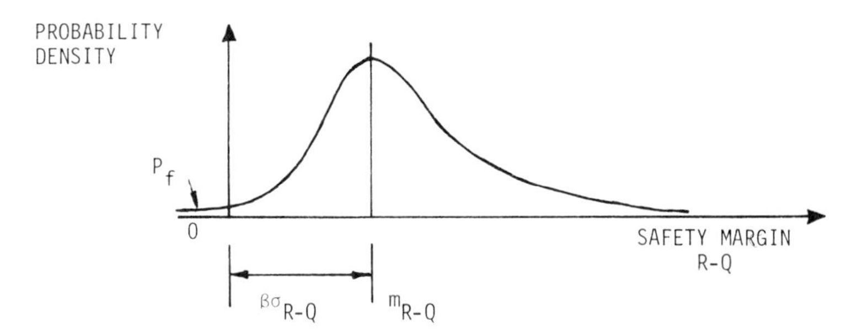

FIG. 8.1. Reliability index model (Cornell, 1969). β = reliability index = m_{R-Q}/σ_{R-Q}.

and the variance reflects the dispersion about the mean. Structural safety is characterised by a relative measure of reliability known as the 'reliability index', β. The reliability index, β, is defined as the ratio of the mean to the standard deviation of safety margin; safety margin is the difference between the resistance, R, of the structural element and the load effect, Q, acting on it. Referring to Fig. 8.1, the reliability index is expressed as the distance between the mean of the safety margin and the point of failure (i.e. where the safety margin equals zero) in terms of the standard deviation of safety margin. Thus

$$\beta = \text{Reliability index} = \frac{m_{R-Q}}{\sigma_{R-Q}} \tag{8.6}$$

in which $m_{(\cdot)}$ and $\sigma_{(\cdot)}$ denote the mean and standard deviation of the random variable $(\cdot)$. If both R and Q have Gaussian distributions, the probability of failure of the structural element can be calculated as $P_f = \Phi(-\beta)$ where $\Phi(\cdot)$ is the standard normal integral. If no distribution assumptions are made, we can only say that a higher value of β indicates higher reliability. Two designs are considered to be consistent if they have the same reliability. Two designs are said to be consistent in the second moment sense if their reliability indices, β, are equal.

The reliability index concept can be extended to design situations where a large number of variables such as loads, material properties, geometrical dimensions and factors representing modelling uncertainties are to be considered. Several second moment formulations have been proposed (Ditlevsen, 1973; Hasofer and Lind, 1974; Rackwitz, 1976). The procedure developed by Rackwitz (1976) follows.

Let $g(x_1, x_2, \ldots, x_n)$ be the mechanical formulation function or 'limit state function' of the reliability problem under study, where $x_1, x_2, \ldots, x_n$ are the basic variables such as material strengths and loads. Failure occurs if and only if $g < 0$. The safety of the failure mode can be assessed, in the process called 'safety checking', by measuring the minimum distance from the mean to any point in the sample space of the structural variables on the surface $g(x_1, \ldots, x_n) = 0$, representing the failure criterion. This point is denoted $\{x_1^*, x_2^*, \ldots, x_n^*\}$. The reliability index is defined as $\beta = m_{g_0}/\sigma_{g_0}$, where g_0 is the linear approximation to $g(\cdot)$:

$$g_0 = g(x_1^*, x_2^*, \ldots, x_n^*) + \sum_{i=1}^{n} \frac{\partial g}{\partial x_i} (x_i - x_i^*) \tag{8.7}$$

in which all the derivatives are evaluated at the point $(x_1^*, \ldots, x_n^*)$. Then

$$m_{g_0} = g(x_1^*, x_2^*, \ldots, x_n^*) + \sum_{i=1}^{n} \frac{\partial g}{\partial x_i} (m_{X_i} - x_i^*) \tag{8.8}$$

$$\sigma_{g_0} = \left[\sum_{i=1}^{n} \left(\frac{\partial g}{\partial x_i} \sigma_{X_i}\right)^2\right]^{1/2} = \sum_{i=1}^{n} \alpha_i \frac{\partial g}{\partial x_i} \sigma_{X_i} \tag{8.9}$$

$$\alpha_i = \frac{\partial g}{\partial x_i} \sigma_{X_i} \left\{\sum \left(\frac{\partial g}{\partial x_j} \sigma_{X_j}\right)^2\right\}^{-1/2} \tag{8.10}$$

By the definition of the reliability index,

$$m_{g_0} - \beta\sigma_{g_0} = 0 \tag{8.11}$$

Substituting eqns (8.8)–(8.10) in eqn (8.11), we obtain

$$g(x_1^*, x_2^*, \ldots, x_n^*) + \sum_{i=1}^{n} \frac{\partial g}{\partial x_i} (m_{X_i} - x_i^* - \alpha_i\beta\sigma_{X_i}) = 0 \tag{8.12}$$

The first term is zero; the second term becomes zero if

$$x_i^* = m_{X_i} - \alpha_i\beta\sigma_{X_i} \tag{8.13}$$

An iterative procedure is used to obtain the design point $\{x_1^*, x_2^*, \ldots, x_n^*\}$:

$$g(x_1^*, x_2^*, \ldots, x_n^*) = g(\ldots, m_{X_i} - \alpha_i\beta\sigma_{X_i}, \ldots) = 0 \tag{8.14}$$

is solved together with the system of equations

$$\alpha_i = \frac{\partial g}{\partial x_i}\Big]_{\mathbf{x}^*}\sigma_{X_i}\left[\sum\left(\frac{\partial g}{\partial x_i}\Big]_{\mathbf{x}^*}\sigma_{X_i}\right)^2\right]^{-1/2} \tag{8.15}$$

The components of $\mathbf{x}^*$ are given by

$$x_i^* = m_{X_i} - \alpha_i\beta\sigma_{X_i} = m_{X_i}(1 - \alpha_i\beta V_{X_i}) \tag{8.16}$$

In eqn (8.16), the term $1 - \alpha_i\beta V_{X_i}$ are the partial safety factors applied on the mean values of the variables; V_{X_i} are the coefficients of variation of variables X_i. The load and resistance factor values to be applied on nominal loads and resistances can be derived from these partial safety factors.

8.1.5 Current Methods in Safety Factor Selection

Safety factors in earlier design codes were selected on the basis of empirical evidence and collective experience of the profession, filtered by the judgement of code writers. The philosophy, rarely stated explicitly, was to establish an upper bound on the load effects and a lower bound on the strength and thus achieve designs that, formally, were absolutely safe. Against the background of the low reliability of the designs of the early industrial era, this philosophy was apparently very successful.

The selection of safety factors such that a given code format meets a specified objective is called 'code calibration'. Modern probabilistic second moment procedures require that the first two statistical moments of the variables be estimated by the code writers; the partial safety factors can then be derived for a given value of the reliability index β. The procedure and the rationale have been laid out in detail by the Canadian Standards Association (1981).

The target reliability index for a structural element may be a value agreed upon by the profession to give the desired level of reliability, or it can be inferred from existing designs. In the latter procedure, values of the reliability index, β, are calculated for a number of structural elements (e.g. simple beams, centrally loaded columns, tension members, high strength bolts and fillet welds) designed according to the design code in service. This phase may be called 'code analysis'. On the basis of this analysis, the code committee selects a set of 'target β-values', B, with which it proceeds to calculate the partial safety factors. This procedure has the advantage of utilising past experience and is based on the precept that the reliabilities inherent in current criteria for 'standard' design situations are known and, if not generally acceptable, modified towards more desirable values. This calibration study was performed in developing the Load and Resistance Factor Design criteria for steel structures (Galambos and Ravindra, 1978). The representative values of β for structural members and connections were judged to be 3·0 and 4·0, respectively. More advanced examples of calibration are described in Section 8.3.

8.1.6 Practical Considerations in Code Calibration

Lind (1976) has outlined the basic steps in developing partial safety factors in a structural design code as summarised below.

The *first step* in the writing of a code is to define the scope or class of structures it is to govern. The scope is a parametered set of structures; the set of parameters is called the data space (described further in Section 8.2). The code is characterised in a mathematical framework as an element of a set called the code format. The numerical constants contained in a code (e.g. load factors, nominal floor live loads and earthquake coefficient) may be considered as variables. As these variables take on various different values, a set of different codes is generated. This set is called the code format of the code; the original code is one of many realisations of the format. Each realisation is characterised by its particular set of values of the parameters, corresponding to a point in the parameter space for the code format.

The *second step* in the writing of a structural code is to define the code objective. For a Level I code (i.e. limit states design code with specified partial safety factors) the objective may be stated at any other level (i.e. maximisation of total expected utility, constant specified reliability over the data set, constant reliability index, or a reliability index of specified variation over the data set, etc.).

The *third step* is the determination of the frequency of occurrence of a particular safety check. Since a code, in general, cannot be both simple and exactly meet the objective, it is necessary to define the most important structural data for which the objective is to be met as closely as possible. For example, if most structural actions at a cross-section are confined within a dead-to-live load ratio of 0·5 to 2·0, it is generally possible to meet the objective more closely over this range than over the hypothetical range of 0 to ∞. The frequency of occurrence is a scalar point function in data space and is called the demand function.

The *fourth step* in the design of a code is to select a measure of closeness between a code and its objective. For example, let B denote the (desired) objective value of the reliability index in a particular safety check, and let β be the actual value produced by the check procedure. The difference $B - \beta$

varies over the data set; for some structural sections (say, 'slender columns') it may be positive while for others (say, 'intermediate columns') it may be negative. Then, the criterion of closeness of the code to the *objective* may be that the expected value of $(B - \beta)^2$ should not exceed a given value, 0·5 for example.

The *fifth step* is the selection of a sequence of trial code formats arranged in order of decreasing simplicity. Even the simplest conceivable objective cannot be met exactly by a level I code except at a practically unacceptable level of complexity (i.e. by having an unmanageable number of code parameters). It is therefore necessary to confine the search to a set of formats that leads to sufficiently simple design procedures. In each format, there exists generally an optimal realisation which comes closest to the objective. With the criterion of closeness, one can select the best of these realisations as the simplest one to meet the criterion.

In Section 8.3 we shall study how these basic steps in code calibration were followed in developing the partial safety factors of the limit states design criteria in the National Building Code of Canada.

8.2 CODE OPTIMISATION

Selection of safety factors in a design code can be viewed as an optimisation process. Selection should be such that optimal designs result for the set of all structures designed using the code. The mathematical framework for codes outlined by Lind (1969) and Ravindra and Lind (1973) offers an objective way of achieving the optimal safety factors by balancing safety and economy as described below.

Safety factors have an influence on the cost, reliability and other aspects of performance of all structures to be designed using the code, and have an ultimate influence on the national economy. Optimisation of an individual structure is not strictly possible because the design has to satisfy the preselected global requirements (i.e. safety factors, minimum reliabilities, etc.) in the code. The designer of any structure has a responsibility to safeguard all persons against loss of life, limb and property that may result from failure, whereas the proper balance between safety and economy can be achieved at the code-writing stage in a block decision.

In the following, a formal mathematical description of code optimisation is given. The function of a structural design code is identified as a mapping of structural data (gross dimensions and layout) into member proportions. A cost of code is defined by assigning a total expected cost function to the mappings considering all the structures to be designed using the code. Code optimisation is formulated as a minimisation of this scalar point function (i.e. total expected cost) and is shown to be a problem in non-linear programming.

8.2.1 Mathematical Structure of Code

The general layout and the principal dimensions of a structure such as span, centre-to-centre distance for beams, and storey heights are specified by the engineer, architect or client. All this information, consisting mainly of numerical values of many independent variables, defines a structure and is called a data set, ω. The totality of all such data sets is called the data space, $\mathbf{\Omega}$. A data set, ω, is called a point in data space.

The points in data space can be classified into different structural schemes. A structural scheme, $\mathbf{S}_i$, is characterised by layout (topology), types of supports, number of storeys in the structure, uses of structure, intended mode of behaviour and method of analysis. This subdivides the data space, $\mathbf{\Omega}$, into subspaces $\mathbf{S}_i$, $i = 1, 2, \ldots$.

The first step in design, once the structural scheme and principal dimensions are given, is to choose the structural material. The material is characterised by a few indices such as specified minimal strength and statistical parameters of its distribution. The designer may also specify the general quality of workmanship and control. The complete set of parameters of optional design variables, called the designer's choices, is denoted by ζ. The totality of all such sets is called the 'designer's choice space', $\mathbf{Z}$; ζ is called a point of $\mathbf{Z}$.

Design, in the narrow sense of selecting member proportions to achieve satisfactory structural performance, means that a point, $\mathbf{m}$, on a member proportion space, $\mathbf{\Lambda}$, is associated with the points ω and ζ:

$$(\omega, \zeta) \rightarrow \mathbf{m} \tag{8.17}$$

Most of the provisions in a code are inequalities. In conjunction with the ambient unit prices, a code defines the member proportions when the objective is structural optimisation in the usual sense of minimising initial cost. The correspondence, eqn (8.17), is a mapping of $\mathbf{\Omega} \times \mathbf{Z}$ into $\mathbf{\Lambda}$ defined by the code and the unit price of the structure. To every design $(\omega, \zeta, \mathbf{m})$ a scalar objective function $c(\omega, \zeta, \mathbf{m})$ can be assigned (as examined subsequently), called the total expected cost and assumed to have the usual properties of a negative utility. A design $(\omega^o, \zeta^o, \mathbf{m}^o)$ is called optimal if no other point exists in $(\omega^o, \zeta^o, \mathbf{m})$ that has a lower value of the cost. A unique optimal design is assumed to exist for each point (ω^o, ζ^o) in a domain of

$\mathbf{\Omega} \times \mathbf{Z}$. Optimal design then defines a single valued mapping of this domain into $\mathbf{\Lambda}$: $\mathbf{m} = \mathbf{m}^{\mathrm{o}}(\omega, \zeta)$.

If a change is made in the design code, it may cause a change in the set of permissible designs for some data points (ω, ζ) and consequently it may change the mapping $\mathbf{m} = \mathbf{m}^{\mathrm{o}}(\omega, \zeta)$. The code is conveniently visualised as a point, $\mathbf{p}$, in the abstract parameter space, $\mathbf{P}$. The point, $\mathbf{p}$, represents both the format (wording) of the code and the set of numerical quantities present in the code (for example, safely factors), called the code parameters, that can be manipulated by the code authority. The influence of the code on the optimum design mapping is expressed symbolically by writing

$$\mathbf{m} = \mathbf{m}^{\mathrm{o}}(\omega, \zeta, \mathbf{p}) \tag{8.18}$$

When a code is revised $(\mathbf{p} \to \mathbf{p}^*)$, some points (ω, ζ) may well be mapped on to proportions with lesser cost while others remain unchanged or increase in cost. Two design codes $\mathbf{p}$ and $\mathbf{p}^*$ can be compared if suitable measures of cost, reliability, etc., can be assigned as scalar functions in parameter space $\mathbf{P}$—if, moreover, an appropriate measure of comparison of the values of these functions can be selected. These measures are considered in the following sections.

8.2.2 Cost of a Code

Two codes are called strictly identical if their mappings are identical. When two codes are not strictly identical, they can be compared by means of a cost function.

The optimal initial cost, $c_{\mathrm{i}}^{\mathrm{o}}$, of a design at point (ω, ζ) under code $\mathbf{p}$ can be readily calculated as

$$c_{\mathrm{i}}^{\mathrm{o}} = c_{\mathrm{i}}^{\mathrm{o}}(\omega, \zeta, \mathbf{p}) = c_{\mathrm{i}}^{\mathrm{o}}[\omega, \zeta, \mathbf{m}^{\mathrm{o}}(\omega, \zeta, \mathbf{p})] \tag{8.19}$$

The initial cost of a code, $C_{\mathrm{I}}^{\mathrm{o}}(\mathbf{p})$, is accordingly

$$C_{\mathrm{I}}^{\mathrm{o}}(\mathbf{p}) = \sum_{\mathbf{S}_i} \int_{\Omega} \int_{\mathbf{Z}} c_{\mathrm{i}}^{\mathrm{o}}(\omega, \zeta, \mathbf{p}) f_{\Omega \mathbf{Z}_i}(\omega, \zeta, \mathbf{p}) \, \mathrm{d}\Omega \, \mathrm{d}\mathbf{Z} \qquad i = 1, 2, \ldots \tag{8.20}$$

in which $f_{\Omega \mathbf{Z}_i}(\omega, \zeta, \mathbf{p}) \, \mathrm{d}\Omega \, \mathrm{d}\mathbf{Z}$ is the joint frequency of occurrence of the data element at point (ω, ζ) in the domain of structural scheme $\mathbf{S}_i$. This frequency is known or must be assumed; its dependence on $\mathbf{p}$ would normally be neglected. The summation and integrations in eqn (8.20) can only be accomplished if the abstract space $\mathbf{\Omega} \times \mathbf{Z}$ can be ordered such that all points can be scanned once and only once in the process. The problem is then to denumerate the set of all structures governed by a code and all designer's options admitted by a code. It has been shown (Ravindra and Lind, 1971) that some common structural schemes, e.g. continuous beams and structural frameworks, can be assigned unique numerical labels. The set of all such structural schemes can be scanned by scanning the set of natural numbers.

The integrand in eqn (8.19) is the total optimal initial cost, $c_{\mathrm{i}}^{\mathrm{o}}(\omega, \zeta, \mathbf{p})$, of all structures represented by the data point, (ω, ζ). In practice, it cannot be expected that all structures at a particular data point will be built (alike) at minimum cost; rather, there will be some scatter such that $\mathbf{m}$ is a stochastic point function in $\mathbf{\Omega} \times \mathbf{Z}$. The expected initial cost will be a value, $c_{\mathrm{i}}(\omega, \zeta, \mathbf{p})$, that is higher than $c_{\mathrm{i}}^{\mathrm{o}}(\omega, \zeta, \mathbf{p})$ by a small percentage that depends mainly on the state of development of the particular structural technology, and on the climate of economic competition in society. The practical cost, c_i, is well defined operationally and can be determined by design and competitive bidding at any data point. The corresponding practical total expected cost function for the code, $C_{\mathrm{I}}(\mathbf{p})$, is given by an equation analogous to eqn (8.20) by dropping the superscript o.

A structure is subjected to random loads, has a stochastic strength, and has other uncertainties associated with it due to imperfections in fabrication, idealisation in analysis, design, etc. The reliability of a structure, according to any of the limit state criteria, can be calculated from the probability distributions of the design variables. Practical calculation of the reliability of a structure, taking into account the various failure modes and loads with their correlations, is not always possible, but some approximate procedures are available (Rackwitz and Peintinger, 1981).

The design variables may be identified as: material strengths, fabrication errors, random loads, errors due to idealisations of strength, structural analyses, etc. (Galambos and Ravindra, 1978). A design variable can be represented in practice by its first n statistical moments as an alternative to its description by a complete probability distribution. A set containing all the variables identified earlier, with their first n moments, is called a set of real physical variables, $\boldsymbol{\theta}$. The totality of all such sets is called the space of real physical variables, Θ. A point in this space represents the set of n moments of every design variable with a frequency of occurrence, $f(\theta)\,\mathrm{d}\Theta$, about it.

A loss can be assigned to every structure, arising out of malfunction such as excessive deflection, structural cracks, and loss of life or limb, loss of goodwill and production. A probability of occurrence is associated with each loss. Each loss is assumed to have an associated socioeconomic equivalent negative utility, called cost of failure herein. This axiom makes the decision a proper subject of economic analysis. The losses are defined to be mutually exclusive events. The total expected cost of failure of a structure, Ec_f, is the sum of the expected costs for each mode of failure, i.e.

$$Ec_f = \sum_{j=1}^{q} c_{fj}(\omega, \zeta, \mathbf{p}) p_{fj}(\omega, \zeta, \mathbf{p}) \tag{8.21}$$

where c_{fj} and p_{fj} are the cost and the probability of failure in mode j, respectively.

The expected cost of failure of a code, $C_F(\mathbf{p})$, can be defined as the sum of the expected costs of failure Ec_f of all structures in the data space designed using this code, with expectations taken over the spaces, $\mathbf{Z}$ and Θ:

$$C_F(\mathbf{p}) = \sum_{\mathbf{S}_i} \int_{\Omega} \int_{\mathbf{Z}} \int_{\Theta} Ec_f f(\boldsymbol{\theta}) \, d\Theta \, f_{\Omega \mathbf{Z}_i}(\omega, \zeta, \mathbf{p}) \, d\Omega \, d\mathbf{Z} \qquad i = 1, 2, \ldots \tag{8.22}$$

The total cost of a code also includes other costs C_0 (of maintenance, demolition, etc.) that are practically independent of $\mathbf{p}$. All costs are discounted to a present value. The codes can now be compared with respect to total cost:

$$C_T(\mathbf{p}) = C_I(\mathbf{p}) + C_F(\mathbf{p}) + C_0 \tag{8.23}$$

8.2.3 Optimal Code

The concept of code cost naturally leads to the notion of an optimal code, defined as a code of least total cost. Because C_T is inherently positive, existence of an optimum code is guaranteed. The space of all possible codes $\mathbf{P}$ is abstract, so that an exhaustive search cannot be made. The attention in the following is restricted to any subset $U\mathbf{P}'$ of $\mathbf{P}$, that is, the sum of a finite number of ordered subspaces $\mathbf{P}'$. Each of these subspaces is a class of codes of similar format, for which the variable, $\mathbf{p}$, may be considered as a vector.

Optimisation of a given format is the minimisation of the scalar point function, $C_T(\mathbf{p})$, in the format parameter space, $\mathbf{P}'$. It is a mathematical programming problem:

Find $\mathbf{p} \in \mathbf{P}'$ such that

$$\begin{aligned} g_k(\mathbf{p}) &= 0 \qquad k = 1, 2, \ldots, K \\ e_j(\mathbf{p}) &\leq 0 \qquad j = 1, 2, \ldots, J \end{aligned} \tag{8.24}$$

minimise $C_T(\mathbf{p})$.

The constraints, $g_k(\mathbf{p})$ and $e_j(\mathbf{p})$, and the objective function, $C_T(\mathbf{p})$, are generally non-linear in $\mathbf{p}$. Non-linear problems of constrained minimisation can be solved by a number of well-known methods. In between codes of different formats, the optimal code is selected by enumerating all formats and by comparing their minimum costs.

8.2.4 Illustration

The following example from Ravindra and Lind (1973) is used to illustrate the code optimisation procedure. The objective was to select load factors on nominal values of dead, wind and snow loads. The structural scheme selected was a one-bay and one-storey steel frame of constant cross-section throughout. The structure was considered to fail when it reached the limit state of plastic collapse. The data variables considered were the span l and height h of the frame. The frequencies of occurrence of different frames used in the example are as shown in Table 8.1.

TABLE 8.1
FREQUENCY OF OCCURRENCE FOR VARIOUS STEEL FRAMES

Height, h (ft)	*Frequency of occurrence of h*	*Length, l (ft)*	*Frequency of occurrence of l*	*Nominal dead load, D_c (lb/ft²)*
10	0·08	15	0·01	25
12	0·70	25	0·20	30
15	0·15	40	0·50	50
20	0·05	60	0·25	80
30	0·02	80	0·04	100

The plastic moment capacity of any cross-section in the frame was estimated to have a mean value $m_R = 1{\cdot}07 M_p$ and a coefficient of variation $V_R = 0{\cdot}13$, where M_p is the nominal plastic moment capacity. The cross-sections were assumed to be partially correlated with each other; a correlation of $\rho = 0{\cdot}80$ was assumed between any two cross-sections. Pertinent data on dead, snow and wind loads were derived from climatological and other information and are shown in Table 8.2.

Collapse of the structural frame under study was considered to occur when any of the possible mechanisms (modes) is formed. The probability of failure of the structure, P_f, under any specified load combination was calculated using the bounds derived by Vanmarcke (1972). Two load combinations were considered: (1) dead load plus lifetime maximum snow load plus annual maximum wind load; and (2) dead load plus annual maximum snow load plus lifetime maximum wind load. The probability of failure of the structure was assumed to be the sum of the failure probabilities for the two combinations.

The members of the frame were assumed to be prismatic and of geometrically similar I-sections. The cross-sectional area of the member was assumed to be related to the nominal moment capacity as $A = 0{\cdot}36M_p^{2/3}$, where M_p is in kip-inches and A is in square inches. The cost of steel was taken as $0·06 per cubic inch. The cost of failure, including loss of life, injury, repair, replacement and loss of prestige, was estimated as

TABLE 8.2
PARAMETERS OF LOADS

Variable	*Nominal value*	*Annual maximum*		*Lifetime maximum*[a]	
		Mean	*Coefficient of variation*	*Mean*	*Coefficient of variation*
Wind load, W (lb/ft^2)	22·8	11·1	0·30	18·6	0·17
Snow load, S (lb/ft^2)	10·0	6·5	0·42	14·5	0·17

[a] Lifetime of the structure is considered to be 50 years.

$300 per square foot of the covered area. The cost was discounted to a present value using a 5% annual discount rate over the life of the structure of 50 years.

The evaluation of optimal load factors was done using the following procedure:

1. Assume a set of load factors γ^*.
2. Design the structure defined by ω for loads $\gamma^*\mathbf{P}_n$ (where $\mathbf{P}_n$ are nominal values of dead load, wind load and snow load).
3. With the assumed statistical parameters of loads and of resistance, calculate $p_f(\gamma^*, \omega)$ for each load combination and add the probabilities to obtain $P_F(\gamma^*, \omega)$.
4. Calculate the initial cost $C_I(\gamma^*, \omega)$ and cost of failure $C_F(\omega)$.
5. Total expected cost $= C_T(\gamma^*, \omega) = C_I(\gamma^*, \omega) + P_F(\gamma^*, \omega)C_F(\omega)$.
6. Repeat steps 2 to 5 for other ω.
7. Calculate $C_T(\gamma^*) = \sum_\omega C_T(\gamma^*, \omega) f(\omega)$.
8. Repeat steps 1 to 7 for other values of γ until $C_T(\gamma)$ is a minimum.

This non-linear mathematical programming problem was solved to yield the optimal load factors of $\gamma_D = 1{\cdot}53$, $\gamma_S = 2{\cdot}75$ and $\gamma_W = 1{\cdot}41$.

8.2.5 Practical Considerations in Code Optimisation

Optimisation of a code in its entirety may be an insurmountable task even with the assistance of modern high-speed computers. The code writers should decide *a priori* which are the more significant code parameters, e.g. load and resistance factors, and thereby reduce the task to a manageable size. Knowledge about the frequencies of occurrence of data and structural schemes is obtained fairly easily from sample surveys in properly chosen localities. In recent applications, these frequencies were obtained through opinions expressed by experienced designers.

The advantage of deriving the safety factors through code optimisation is that any new information (e.g. load data and frequencies of data) and research results (e.g. reliability models and costs of failure) may be processed to update generally the numerical values of the safety factors in a chosen code format. This is further elaborated in Section 8.3. The output of code optimisation could differ depending on the level of code under development. In a Level I code, it is a set of load and resistance factors specified in selected load combinations. A Level II code requires the optimal values of reliability indices for different structural elements. A Level III code requires the optimal values of probabilities of failure for different structural schemes.

Applications of code optimisation to date have not made explicit calculations of expected costs of codes. But the selection of safety factors is being done by considering the frequencies of occurrence of different design situations.

8.3 SELECTION OF SAFETY FACTORS

In recent years, some major structural design codes have been drafted based on limit states design philosophy. Examples include the National Building Code of Canada (Siu *et al.*, 1975), the Ontario Highway Bridge Code (Lind and Nowak, 1978) and the American National Standards Institute's Loading Standard (Ellingwood *et al.*, 1980). The load and resistance factors in these codes have been selected using the code optimisation concepts of data space and measure of closeness as discussed in Section 8.1. In the following, we describe some of these applications.

8.3.1 National Building Code of Canada Limit States Design Criterion (Siu *et al.*, 1975)

The limit states design criterion in the National Building Code of Canada for all materials and types of construction is presented in the following format:

$$\phi R_n \geq \gamma_D D_n + \psi(\gamma_L L_n + \gamma_W W_n + \gamma_T T_n) \qquad (8.25)$$

The selection of code parameters ϕ, γ_D, γ_L, γ_W, ... and ψ was based on a calibration to existing design codes for cold-formed steel, hot-rolled steel, reinforced concrete and wood. The objective was to have common load factors for all these materials of construction; the resistance factors would differ based on material and limit state. The statistical data on loads and limit state resistances were collected and are listed in Table 8.3. The nominal value of a load was taken to be the value specified by the code authorities, and the nominal resistance was assumed to be the value

TABLE 8.3
STATISTICAL DATA

Variable	*Mean/nominal*	*Coefficient of variation*
Loads		
Dead load	1·00	0·07
Live load	0·70	0·30
Wind load	0·80	0·25
Materials—Limit states		
Light-gauge steel		
Tension and flexure	1·20	0·14
Hot-rolled steel		
Tension and flexure	1·10	0·13
Compression	1·20	0·15
Reinforced concrete		
Flexure	1·14	0·15
Compression	1·14	0·16
Shear	1·10	0·21
Timber		
Tension and flexure	1·31	0·16
Compression parallel to grain	1·36	0·18
Compression perpendicular to grain	1·71	0·28
Shear	1·26	0·14
Buckling	1·48	0·22

From Siu *et al.* (1975). Courtesy American Society of Civil Engineers.

predicted by theoretical or empirical formulae used in conjunction with the design standard.

The data space consisting of different structures, different locations in the structure and geographical location was characterised in this study by two variables: live to dead load ratio p and wind to dead load ratio q. The frequencies of occurrence of different design situations were estimated on the basis of judgement by several individuals, and are presented as weighting factors for combinations of p and q as shown in Table 8.4. The data space would vary with the technology (i.e. material) and the individual

TABLE 8.4

WEIGHTING FACTORS FOR VARIOUS VALUES OF p AND q

q	p					
	0·0	*0·5*	*1·0*	*2·0*	*3·0*	*4·0*
0·0	0	5	10	8	2	2
0·5	4	6	12	10	8	2
1·0	8	10	15	20	10	2
2·0	6	4	8	4	2	1
3·0	2	2	1	1	0	0

From Siu *et al.* (1975). Courtesy American Society of Civil Engineers.

limit states and should be properly represented in the calibration by the code committee on the basis of usage. However, in this study only one set of design situations and their frequencies of occurrence was assumed for all technologies and limit states.

The calibration procedure used in this study is:

(a) For each limit state for the structural material (e.g. hot-rolled steel column compression failure), the reliability index values implied in the existing code for all design situations were calculated. The value of the reliability index β was determined using the following equation:

$$\beta = \frac{\ln\left(\dfrac{m_R}{m_Q}\right)}{\sqrt{(V_R^2 + V_Q^2)}} \tag{8.26}$$

in which m_R and m_Q are, respectively, the mean resistance and mean total load effect, and V_R and V_Q are the coefficients of variation. Table 8.5 shows the implied reliability index values for hot-rolled steel columns under compression failure.

A weighted average value of the reliability index, β_{avg}, was found for the structural material for the limit state under consideration:

$$\beta_{avg} = \frac{\sum f_i \beta_i}{\sum f_i} \tag{8.27}$$

where f_i is the weighting factor based on the frequency of occurrence of the specific design situation (i.e. combination of p and q values in Table 8.4).

TABLE 8.5

EXAMPLE β-VALUES[a] IN EXISTING DESIGN STANDARDS[b]

q	p					
	0·00	*0·50*	*1·00*	*2·00*	*3·00*	*4·00*
0·00	5·04	5·32	5·02	5·02	4·30	4·14
0·50	5·21	4·64	5·00	4·55	4·30	4·14
1·00	4·93	4·93	4·83	4·55	4·30	4·14
2·00	4·52	4·52	4·52	4·83	4·69	4·32
3·00	4·29	4·29	4·29	4·80	4·79	4·68

[a] Weighted average 4·67.
[b] Factor of safety in existing design standard = 1·92.

(b) From an analysis of the β_{avg} values for different materials under different limit states, representative 'target' β_0 values were selected for different limit states as constants for all structural members, e.g. $\beta_0 = 4{\cdot}00$ for yielding in tension and flexure, $\beta_0 = 4{\cdot}75$ for compression and buckling failures, and $\beta_0 = 4{\cdot}25$ for shear failures.

(c) The evaluation of ϕ, γ_D, γ_L, ... was carried out by using the code optimisation procedure. For a selected set of ϕ, γ_D, γ_L, ... and for a given material under a particular limit state, the implied value of the reliability index, denoted b, was determined for the new code at a chosen data point. An objective function representing the measure of closeness was formulated as

$$Z = \sum_{\text{matls}} \sum_{\substack{\text{limit} \\ \text{space}}} \sum_{\substack{\text{data} \\ \text{space}}} (\beta_0 - b)^2 f_i \tag{8.28}$$

where β_0 is the target reliability index for a particular limit state and material. The minimisation of Z resulted in the following optimal values of ϕ, γ_D, γ_L, γ_W and ψ:

$$\phi R_n \geq \begin{cases} 1{\cdot}25D_n + 1{\cdot}50L_n \\ 1{\cdot}25D_n + 0{\cdot}70(1{\cdot}50L_n + 1{\cdot}40W_n) \end{cases}$$

The resistance factors for different technologies were obtained as:

Cold-formed steel:	Yielding	$\phi_1 = 0{\cdot}90$
Hot-rolled steel:	Yielding	$\phi_2 = 0{\cdot}85$
	Compression	$\phi_3 = 0{\cdot}74$
Reinforced concrete:	Flexure	$\phi_4 = 0{\cdot}83$
	Compression	$\phi_5 = 0{\cdot}68$
	Shear	$\phi_6 = 0{\cdot}64$
Wood:	Flexure	$\phi_7 = 0{\cdot}92$
	Compression parallel to grain	$\phi_8 = 0{\cdot}76$
	Compression perpendicular to grain	$\phi_9 = 0{\cdot}64$
	Shear	$\phi_{10} = 0{\cdot}90$
	Buckling	$\phi_{11} = 0{\cdot}70$

The advantages of this calibration scheme are (1) it is adaptable to any safety-evaluating rationale (e.g. advanced first-order, second moment reliability index proposed by Rackwitz and Peintinger (1981), design for specified reliabilities, etc.), (2) it can accommodate many design standards simultaneously to yield common loading criteria, and (3) the method is flexible in defining the objective of calibration.

8.3.2 American National Standards Institute's Loading Standards

The objective of the calibration study conducted by Ellingwood *et al.* (1980) was to develop common loading criteria for different construction materials. The following limit states design equation was used:

$$\phi R_n \geq \gamma_D D_n + \gamma_Q Q_n + \sum \gamma_j Q_{nj} \qquad (8.29)$$

where $\gamma_Q Q_n$ = factored principal variable load, and $\gamma_j Q_{nj}$ = factored arbitrary point-in-time loads. The load combination approach suggested by Turkstra (1972) was used in formulating eqn (8.29).

Based on an analysis of implied reliabilities in current material design standards for steel, reinforced concrete, glulam and masonry, the following

TABLE 8.6
WEIGHTS FOR $D + L$ AND $D + S$ LOAD COMBINATIONS

Material	*Combination*	L_n/D_n, S_n/D_n						
		0·25	*0·50*	*1·0*	*1·5*	*2·0*	*3·0*	*5·0*
Steel	$D + L$, $D + S$	0·0	0·10	0·20	0·25	0·35	0·07	0·03
Reinforced concrete	$D + L$	0·10	0·45	0·30	0·10	0·05	0·0	0·0
Cold-formed steel	$D + S$	0·30	0·40	0·20	0·05	0·05	0·0	0·0
Aluminium	$D + L$, $D + S$	0·0	0·0	0·06	0·17	0·22	0·33	0·22
Glulam	$D + L$	0·0	0·05	0·26	0·26	0·26	0·12	0·05
	$D + S$	0·0	0·02	0·16	0·32	0·32	0·18	0·0
Masonry	$D + L$, $D + S$	0·36	0·36	0·20	0·06	0·02	0·0	0·0

target β_0 values were chosen for deriving the load criteria: for $D + L$ and $D + S$, $\beta_0 = 3{\cdot}0$; for $D + L + W$, $\beta_0 = 2{\cdot}5$; for $D + L + E$, $\beta_0 = 2{\cdot}0$; and for counteracting loads, $\beta_0 = 2{\cdot}0$. The reliability index β was evaluated using the procedure of Rackwitz (1976) as described in Section 8.1.

The load factors in eqn (8.29) were derived by minimising the following objective function:

$$Z(\phi, \gamma_Q, \gamma_j) = \sum_i (R_{ni}(\beta_0) - R_{n1_i})^2 f_i \tag{8.30}$$

where $R_{n_i}(\beta_0)$ is the nominal resistance calculated using the reliability index β_0 for the ith load situation, and R_{n1_i} is the nominal resistance calculated using a selected set of constant load factors. f_i is the relative weight assigned to the ith load situation. Table 8.6 shows the weights assigned on the basis of relative frequency of different load situations and construction materials for gravity load combinations. For combinations involving wind and earthquake loads, it was assumed that values of W_n/D_n and E_n/D_n of 0·5, 1·0, 3·0 and 5·0 were equally likely.

The following load combinations applicable for all the construction materials included in this study were obtained:

$$\phi R_n \geq \begin{cases} 1{\cdot}4D_n \\ 1{\cdot}2D_n + 1{\cdot}6L_n \\ 1{\cdot}2D_n + 1{\cdot}6L_n + (0{\cdot}5S_n \text{ or } 0{\cdot}8W_n) \\ 1{\cdot}2D_n + 1{\cdot}3W_n + 0{\cdot}5S_n \\ 1{\cdot}2D_n + 1{\cdot}5E_n + (0{\cdot}5S_n \text{ or } 0{\cdot}2S_n) \\ 0{\cdot}9D_n - (1{\cdot}3W_n \text{ or } 1{\cdot}5E_n) \end{cases} \tag{8.31}$$

The specification writing groups may determine the resistance factors ϕ for a given target β_0 level that are consistent with the above load combinations.

8.3.3 Load Combination Methodology Development for Nuclear Components

(*a*) *Format*

Some preliminary studies towards developing a load combination methodology for nuclear components (e.g. vessels, piping system, pumps and valves) were conducted by the Lawrence Livermore National Laboratory (Schwartz *et al.*, 1981). In the following, the salient features of this effort will be discussed, highlighting the differences between this and other code calibration studies.

The design format chosen for load combinations is

$$\phi R_n \geq \gamma_1 c_1 X_{n1} + \gamma_2 c_2 X_{n2} + \gamma_3 c_3 X_{n3} + \cdots \tag{8.32}$$

where $c_1, c_2, c_3, \ldots$ are influence coefficients and $X_{n1}, X_{n2}, X_{n3}, \ldots$ are nominal loads. The values of $\phi, \gamma_1, \gamma_2, \ldots$ are selected such that the resulting component designs have limit state probabilities that are less than the target value.

(*b*) *Objective*

In contrast to other code calibration studies discussed earlier, the objective in this project was to obtain load combinations for components which would have constant specified reliabilities rather than constant reliability indices. This illustrates how a Level I code can be derived from Level III code considerations.

Nuclear components are subjected to a number of operating and extreme loads; they are sustained, transient or dynamic. Most of them have random time-varying occurrences, durations and magnitudes. Calculation of the limit state probability of a component subjected to multiple random loads should consider the combination of dynamic responses. The limit state probability P_F is expressed as

$$P_F = \int_0^{\infty} [1 - F_{Y_t}(r)] f_R(r)\,dr \tag{8.33}$$

where $F_{Y_t}(r)$ is the distribution of the total combined response. The total combined response is

$$Y_t = Y_c + Y_{int} \tag{8.34}$$

where Y_c = continuous load response and Y_{int} = intermittent load response. The total responses from combinations of continuous loads and of intermittent loads were calculated using the 'point crossing formula' (Larrabee and Cornell, 1981). All possible combinations of loads were considered in the calculation of the probability distribution of total combined response using a load event tree (Winterstein, 1980).

(*c*) *Data Space*
It is intended to categorise the components into vessels, concrete reactor vessels, concrete containments, piping systems, pumps, valves, core support structures and storage tanks. Within each category of component (e.g. piping systems) a further subdivision may be necessary based on safety function (e.g. Class 1, 2 and 3). Yet another subdivision may be into subclasses such as main steam lines, SRV lines, reactor coolant piping and auxiliary feedwater piping. For each of these subclasses, a set of design load combinations is developed corresponding to a set of target limit state probabilities. The frequencies of occurrence of different subclasses may be obtained from surveys of operating nuclear power plants.

(*d*) *Load and Resistance Factors*
Optimal load and resistance factors for a selected set of load combinations are derived by minimising the following objective function:

$$Z(\phi, \gamma) = \sum_{\omega \in \Omega} f(\omega) \left[\frac{\log P_F(\omega) - \log P_T}{\log P_T} \right]^2 \tag{8.35}$$

where P_T = target limit state probability and $f(\omega)$ = frequency of occurrence of data, ω. As an illustration, the methodology was applied to a piping system subject to pressure, hydraulic transient, thermal and earthquake loads. The limit states considered were pipe rupture, restraint buckling and unacceptable valve acceleration. The resistance factors were also evaluated for load combinations consisting of unit load factors.

8.4 DISCUSSION

Code calibration and design code development studies over the last decade have provided many insights into the functioning of a design code. It is generally understood that code design rules may be maintained simple and deterministic but should be based on probabilistic concepts. The role of a code committee in processing the probabilistic input into a set of deterministic design rules is emphasised. In discharging this role, the code authorities have the responsibility of conducting the probabilistic analysis and calibration studies that may assist in their judgement. The basis for the selection of safety factors in a code will need to be disclosed in terms of the objective, target reliability indices or reliabilities, data space and demand function. It is also shown that a common basis for design with different materials can be established and that common loading criteria can be derived for different technologies. Several aspects of code calibration and design code development requiring further discussion are as follows.

8.4.1 Selection of Code Formats

The selection of a code format consists of the choices of safety-checking equation (e.g. eqn (8.4)), the specific load combinations and the specific limit states to be considered. Loads are random, time-varying phenomena; their occurrences, durations and magnitudes are also random. How loads combine and how the load effects (responses) combine are the topics of research in stochastic processes. Based on the collective experience of the code committee and the potential for acceptance by designers, a small set of significant load combinations is postulated with the goal of seeking the maximum load effect acting on the structural element during its lifetime. However, in evaluating the reliability (or index) of a structural element, the complete spectrum of load combinations may have to be considered (Section 8.3.3).

Although most design codes are in the process of adopting some form of LRFD format, there may be situations were the designers do not want any change from existing practice. For example, the ASME Boiler and Pressure Vessel Code contains service limits (equivalently allowable stresses) for different loading conditions. The loads in the combinations are not factored. Since the ASME component designers are so much accustomed to this format, it would not be feasible to force an LRFD format into this practice. A practical solution may be to maintain the present format but revise the service limits based on probabilistic calibration studies.

8.4.2 Data Space

In most calibration studies to date, the partial safety factors (or ϕ and γ) were derived by considering a number of design situations that are representative of future use of the code. The design situations included different structural elements (beams, columns, etc.), locations in the structure, different structures and different geographical locations. They followed the existing practice of designing cross-sections rather than structural systems. The load effect at a cross-section was assumed to be fully defined by knowing the influence coefficients for different loads. The task was further simplified to the estimation of the range of variation of a load effect to the known load effect (e.g. S_n/D_n, L_n/D_n, etc.). Where such simplification is not possible, the development of partial safety factors requires, in general, that a representative sample of components (or structures) be designed and analysed. This general procedure may prove to be impracticable in some instances (e.g. piping systems, continuous beams). If we consider a piping system as an example of a component, current design practice is as follows. A piping thickness is first selected on the basis of internal pressure. The pipe supports and their locations are chosen such that there is no single point in the piping where the ASME service limits are exceeded. Satisfactory design of a piping system is achieved after a number (5–10) of iterations of stress analysis for different arrangements of pipe supports. This system is then analysed for the limit state probabilities when subjected to random, time-varying loads. Since many representative components are to be studied as part of the calibration study, this direct procedure of design and analysis requires many costly static and dynamic structural and piping system analyses. Furthermore, in generating the optimal load and resistance values that meet the target in selected load combinations, the limit state probabilities are obtained only after a large number of trials, each trial involving the study of component designs that require several iterations of design and analysis. Based on the concept of influence coefficients, a simulation scheme has been developed (Schwartz *et al.*, 1981) for obtaining a multitude of characterisations of component systems. The responses at different cross-sections (nodes) of the piping system are obtained by simulations using existing design information.

8.4.3 Reliability Analysis

Target reliability indices in a Level I or II code are obtained by calibration to past experience. In the development of LRFD criteria for nuclear components, the target limit state probabilities were specified. In theory, the component (e.g. piping system) reliabilities should be derived by an allocation procedure such that the system consisting of these components will have an acceptable (desired) reliability. This allocation problem is rather complex; some preliminary thoughts are reported by Schwartz *et al.* (1981).

Calculation of the reliability of structural systems considering the correlations between loads and between member resistances, and the appropriate probability distributions for loads and resistances, is indeed a difficult task. Some approximations are available in the literature (Rackwitz and Peintinger, 1981). The merit of deriving load and resistance factors in a selected code format for a target reliability is that it is easy to revise the factors as new developments take place in reliability analysis. The change is relatively smooth.

8.4.4 Future Research Topics

1. Data on frequency of occurrence of different design situations need to be collected in addition to the statistical data on loads and resistances.
2. Sensitivity studies need to be conducted to assess the influences of assumed frequencies of data, probability distributions of load and resistance variables, failure costs and bounds on system reliability.
3. Most code calibration studies have emphasised the ultimate limit states; probability models are needed on serviceability limit states (Turkstra and Reid, 1981).
4. The need and method for incorporating the effects of design and construction errors deserve detailed investigation.

8.5 CONCLUSIONS

From this review of trends in safety factor optimisation, the following conclusions may be drawn:

1. Design codes can be made to retain their deterministic appearance by specifying safety factors (i.e. load and resistance factors) in selected load combinations; the values of these safety factors can be derived on probabilistic analysis. The LRFD code can be developed with any level of code (e.g. specified reliability indices, specified reliabilities and code optimisation) in the background.
2. Design codes for different technologies can have a common basis; they can also have common loading criteria.
3. The safety factors in a code should be derived to meet a specified probability target (β_0, P_T, etc.) over a spectrum of future designs by that code.
4. Calibration studies have proved to be invaluable in the code decision-making process.

REFERENCES

ACI (1977) *Building Code Requirements for Reinforced Concrete*, ACI 318-77, ACI Publications, Detroit, Mich.

AISC (1969) *Specification for the Design, Fabrication and Erection of Structural Steel for Buildings*, American Institute of Steel Construction (Feb.).

ASSOCIATE COMMITTEE ON THE NATIONAL BUILDING CODE (1980) *National Building Code of Canada*, National Research Council of Canada, Ottawa.

CANADIAN STANDARDS ASSOCIATION (1981) Guidelines for the development of limit states design. CSA Special Publication S408-1981, Rexdale, Ontario.

COMITÉ EURO-INTERNATIONAL DU BÉTON (1976) First order reliability concepts for design codes. CEB *Bulletin d'Information*, No. 112, Munich (July).

CORNELL, C. A. (1969) Structural safety specifications based on second-moment reliability analysis. *Final Report, Symposium on Concepts of Safety of Structures and Methods of Design, London*, International Association for Bridge and Structural Engineering, Zürich, pp. 235–46.

DITLEVSEN, O. (1973) Structural reliability and the invariance problem. Report No. 22, Solid Mechanics Division, University of Waterloo, Ontario.

ELLINGWOOD, B., GALAMBOS, T. V., MACGREGOR, J. G. and CORNELL, C. A. (1980) Development of a probability based load criterion for American National Standard A58. National Bureau of Standards Publication 577 (June).

GALAMBOS, T. V. and RAVINDRA, M. K. (1978) Load and resistance factor design for steel. *Journal of the Structural Division, Proc. ASCE*, **104**(ST9), Proc. Paper 14008, 1335–59.

HASOFER, A. M. and LIND, N. C. (1974) Exact and invariant second-moment code format. *Journal of the Engineering Mechanics Division, Proc. ASCE*, **100**(EM1), Proc. Paper 10376, 111–21.

LARRABEE, R. D. and CORNELL, C. A. (1981) Combination of various load processes. *Journal of the Structural Division, Proc. ASCE*, **107**(ST1), 223–39.

LIND, N. C. (1969) Deterministic formats for the probabilistic design of structures. In *An Introduction to Structural Optimization*, ed. M. A. Cohn, Solid Mechanics Study No. 1, University of Waterloo, Ontario, pp. 129–42.

LIND, N. C. (1976) Applications to design of Level I codes. In: First order reliability concepts for design codes, CEB *Bulletin d'Information*, No. 112, Munich (July).

LIND, N. C. and NOWAK, A. (1978) Risk analysis procedure II. Dept of Civil Engineering, University of Waterloo, Ontario (Mar.).

MACGREGOR, J. G. (1976) Safety and limit states design for reinforced concrete. *Canadian Journal of Civil Engineering*, **3**(4), 484–513.

MATOUSEK, M. (1977) Outcomings of a survey on 800 construction failures. *Proceedings, IABSE Colloquium on Inspection and Quality Control*, Institute of Structural Engineering, Swiss Federal Institute of Technology, Zürich.

MAYER, M. (1926) *Die Sicherheit der Bauwerke*, Julius Springer, Berlin.

NORDIC COMMITTEE ON BUILDING REGULATIONS (1978) Recommendation for loading and safety regulations for structural design. NKB Report No. 36, Copenhagen (Nov.).

NOWAK, A. S. and LIND, N. C. (1979) Practical bridge code calibration. *Journal of the Structural Division, Proc. ASCE*, **105**(ST12), Proc. Paper 15061, 2497–510.

RACKWITZ, R. (1976) Practical probabilistic approach to design. In: First order reliability concepts for design codes, CEB *Bulletin d'Information*, No. 112, Munich (July), pp. 13–72.

RACKWITZ, R. and PEINTINGER, B. (1981) General structural system reliability. Presented at the Enlarged Meeting of CEB-Commission 11, Pavia (5–6 Oct.).

RAVINDRA, M. K. and LIND, N. C. (1971) Optimization of a structural code. *Proceedings, Conference on Application of Statistics and Probability to Soil and Structural Engineering*, University of Hong Kong.

RAVINDRA, M. K. and LIND, N. C. (1973) Theory of structural code optimization. *Journal of the Structural Division, Proc. ASCE*, **99**(ST7), 1541–54.

RAVINDRA, M. K., LIND, N. C. and SIU, W. W. (1974) Illustrations of reliability-based design. *Journal of the Structural Division, Proc. ASCE*, **100**(ST9), Proc. Paper 10779, 1789–812.

SCHWARTZ, M. W., RAVINDRA, M. K., CORNELL, C. A. and CHOU, C. K. (1981) Load combination methodology development. Load Combination Program: Project II Final Report, NUREG/CR-2087, UCRL-53025, Lawrence Livermore Laboratory, Calif. (July).

SIU, W. W., PARIMI, S. R. and LIND, N. C. (1975) Practical approach to code calibration. *Journal of the Structural Division, Proc. ASCE*, **101**(ST7), Proc. Paper 11404, 1469–80.

TURKSTRA, C. J. (1972) Theory of structural design decisions. Solid Mechanics Study No. 2, University of Waterloo, Ontario.

TURKSTRA, C. J. and REID, S. G. (1981) Structural design for serviceability. *Proceedings of the Symposium on Probabilistic Methods in Structural Engineering*, St Louis, Mo. (26–27 Oct.), ASCE, pp. 81–101.

VANMARCKE, E. H. (1972) Matrix formulation of reliability analysis and reliability-based design. Presented at the National Symposium on Computerized Structural Analysis and Design, Washington, DC (Mar.).

WINTERSTEIN, S. R. (1980) Stochastic dynamic response combinations. MS thesis, Dept of Civil Engineering, Massachusetts Institute of Technology, Cambridge, Mass. (May).

INDEX

Accuracy, 21, 22
ACI codes, 210
AISC codes, 83–5, 187–93, 198, 203–4, 209
AISI specifications, 169–70, 178, 180
Allowable stress designs, 187, 191–3
American design methods, proposed, 91–2
American National Standards Institute, 228–30
Amplification factors, 190, 194, 196, 198
Analytical methods, 50–6, 145–54, 165–79, 187–93
Antisymmetrical moment patterns, 21
ASME codes, 232
Australian design methods, 85–6
Axial loading, 111, 147–8, 155, 187

Beam
  parameters, 8, 11, 16, 194, 196
  strengths, effect of buckling, 186
  theory approximations, 102, 105–6, 138
Bending analysis, 50–1
Biaxial loading, 191, 193, 198–200
Bimetallic strips, 47
Boundaries, yielded regions, 45
Bowing, 28–30, 53
Braced beams, 63–4, 161–83, 189
Bracing, 162–4, 166
British Standards
  bridge design (BS 5400: Part 3: 1982), 90–1
  building (BS 5950), 91
  steel beams (BS 449: 1969), 81–3; (BS 153: 1972), 81
Buckling
  affecting beam strengths, 186
  analysis, 51–6
  perforated beams, 126–8
  strengths, 37–8
Buildings, 91, 97, 161–5, 181–3, 191, 196

Calibration studies, 215–17, 226, 228, 232, 234
Canadian building codes, 210, 211, 217, 224–8
Canadian Standards Association, 212, 215
Cantilevers, 15–17, 63, 82, 109
Castellated beams, 108–9
Central loading, 12, 49, 53–6, 63, 102, 107, 150–1, 186–7
Channel beams, 163–5
Circular web openings, 100–7, 113, 116
Codes
  calibration, 216–17, 232, 234

Codes—*contd.*
cost, 217, 219–21
formats, 211–12, 230, 232
mathematical structure, 218–19
optimisation, 217–24
revision, 219
safety factors, 209–11
Cold-working, 41
Collapse analysis, 156–8
Complex variable methods, 100–1, 103
Component, categorisation, 231, 233
Composite beams, 118–22
Compressive strengths, concrete slabs, 121
Compressive stresses, 82
Computer-based methods, 52, 60, 86, 168, 205–6
Concrete slabs, 118–19, 121–2
Continuity effects, 19–21, 63–4
Continuous beams, 17–21, 64–6, 79–80
Cooling, stresses, 41–2
Costs, 97, 217, 219–21
Critical conditions, 144, 146, 147–8
Critical elastic stresses, 82, 83, 85
Critical loads, 13, 16, 18, 23, 140–1, 152
Critical moments, 21, 23, 151–2, 154
Cross-section analysis, 40, 43–5
Cuk and Trahair method, 21, 22
Curvatures, 43–5, 50, 139

Data space, 216, 218–19, 226, 231, 232–3
Deep beams, 15
Deflections
amplitudes, 174
components, 170
effects, 175
perforated beams, 106–7
prediction, 174–5, 181
Design
aids, 106, 128–9, 203
methods, 80–92, 187–93
requirements, 3, 73, 97, 128–9
Determinate beams, 50, 63, 78–9, 139
Diaphragms, defined, 162
Dimensionless moments, 77–9
plotted, 18, 29, 30, 38, 45, 48, 54–5, 57, 59, 60, 62, 64, 89
Displacement functions, 174
Displacements
components, 107, 167
effects, 138–9, 150–2
Dux and Kitipornchai method, 20, 22

Eaves, 165
Eccentric loading, 61, 74
Eccentric web openings, 117–18, 120
Edge stresses, 99, 102
Effective lengths, 13–14, 17, 140
charts, 19, 202–3
factors, 14, 19, 195–7, 202–3
Elastic behaviour, perforated beams, 100–7
Elastic buckling, 3–30
Elastic line models, 39–40
Elasticity theory, 100–3
End braces, 163–4
End moments, unequal, 9–12, 61–3, 79–80, 190
End restraints, 11–17, 194–5
axial straining, 139
condition, 23
parameters, 13–14
Energy equations, 141, 143–5, 173–4
Equilibrium equations, 143–4, 187
Equivalent end moments, 61
Equivalent uniform moment factors, 9–12, 16, 19, 85, 89, 92
Errors, effect, 210–11, 234
Euler theory, 23
Event trees, 231
Experimental data
compared, 76, 113, 158
reviewed, 74–80, 181

Failure
costs, 221
criteria, 157–8, 178–9, 183
envelopes, 140–1, 157–8
modes, 125, 127, 179, 186, 194–5
sudden, 3, 5

Fatigue loading, 99, 106
Fillets, rolled sections, 7
Finite difference methods, 53–4, 55, 103
Finite element analysis, 22, 54, 55, 56, 60, 104–5, 109, 140–1, 152, 154–5
Finite integral methods, 54–5
Flame-cutting, 42, 58
Flanged beams, 25, 26
Flanges
  loading, 12, 55
  shear forces, 171–3
  stresses, 58–9
  yielding, 44
Flexural rigidities, 7–8
Flooring, 97, 107, 118

Galerkin method, 167, 168
Geometrically non-linear analysis, 144–5, 154–6
Geometries, 24–5, 40, 110, 172
  web openings, 110, 117
Gravity loads, 197–8

Hinge collapse mechanisms, 108, 118, 120
Hot-rolled beams, 38–9, 41–2, 53–64, 76–9, 82
Hot-rolled sections, 87

I-beams
  elastic buckling, 3–30
  inelastic buckling, 38–66
  instability, 147–51, 156–7
  web openings, 97–129
I-sections, detailed, 4, 6
Idealised behaviour, 169–71
Imperfection parameters, 87–8, 91
Imperfections, 27–8, 39, 64, 73–4, 78, 80, 143, 155, 156, 157
In-plane behaviour, 200–1
In-plane bending analysis, 50–1
In-plane bending moments, 44–5
In-plane deflections, effect, 21–3
Incomplete torsional restraint, 14–15
Incremental analysis, 144–5
Inelastic buckling, 37–66
Inelastic rigidities, 47–8
Initially deformed beams, 27–30
Inspection, 210–11
Instability
  analysis, 145–54, 195
  modes, 137–8
Interaction
  buckling, 17
  diagrams, 109, 112, 115–16, 119, 127
    design aids, 106
  equations, 188–95
Iterative procedures, 52, 177, 187, 214–15, 223, 233

K-factors, 8, 11, 16, 194, 196
Kitipornchai and Trahair calculations, 53–5

Large beams, 3–4
Lateral buckling, defined, 4–5
Lateral slenderness ratios, 91
LeMessurier's method, 195, 197
Length effects, 8
Limit state designs (LSD), 90, 91, 188, 224–9
Limiting stresses, 30, 81
Line models, 39–40, 46–9, 52
Linear interaction equations, 194, 199, 201
Lipped channels, 170–3, 181–3
Load
  combinations, 222–3, 229
  deflection curves, 75, 139, 155, 180
  factors, 223, 225, 231
  notation, 167
  parameters, 223
  resistance factor designs (LRFD), 90, 91, 188, 216, 232–4
Lower bound approaches, 18–19, 22, 86–9, 108, 140–1
Lower bound strengths, 86–7, 89

Manufacturers, metal buildings, 181–3
Mapping, 100, 102, 217–19
Materials, properties, 41, 225, 226, 228–9
Maximum permissible stresses, 82, 84–5
Michell analysis, 4
Moment capacities, 222
Moment-to-shear ratios, 109–15
Moments
  beam column bending, 176
  capacity, 80
  diagrams, 9–11, 22–3, 49
  displacement curves, 156
  distribution, 21, 49–51, 59–63, 88–9
  gradients, 9–11, 60–3, 85
  in-plane bending, 44–5
  inertia, 175
  patterns, 9–11, 22–3
  ratios, 9–11, 22, 60–3, 79–80, 190
Monosymmetric beams, buckling, 24–6
Monosymmetry parameters, 24–5, 47
Multiple column curves, 205
Multiple web openings, 104, 125–7
Multi-storey buildings, 191, 196
Muskhelishvili approach, 100–1

National Building Code of Canada, 210, 211, 217, 224–8
Nethercot and Trahair method, 19, 22, 63
Non-linear interaction equations, 197–200
Non-linear strains, thin-walled beams, 142–3
Non-uniform beams, buckling, 26–7
Non-uniform bending, buckling, 8
Non-uniform yielding, 49
Nuclear components, 230–1

Occurrence, frequency, 216, 222, 234
Out-of-plane behaviour, 200–1
Out-of-plane buckling analysis, 51–6

P–Δ effects, 190–1, 194–5, 197, 200
Perry curves, 140
Perry–Robertson method, 81, 140
Pin-ended columns, 13, 155
Plastic analysis, 188–91
Plastic behaviour, perforated beams, 107–28
Plastic collapse mechanisms, 64, 76, 108
Plate
  assembly models, 52–3
  element models, 39
  girders, 81–2, 126–8
  theory approximations, 138
Point crossing formula, 231
Point-matching, 103
Poisson's ratio, 41
Post-buckling analysis, 23–4
Prandtl analysis, 4
Precambering, 21
Probabilistic design codes, 210, 218–23, 230–1
Purlins
  defined, 161–2
  reversed, 165
  wind uplift loading, 168–79

Quasi-elastic models, 51

Rackwitz procedure, 213–15, 229
Real beams, 73–5
Rectangular web openings, 100–1, 116
Redundant actions, 50–1
Reinforcement
  types, 98–9, 104, 125
  web openings, 122–5
Reliability
  analysis, 233–4
  index, 213, 226–7
Residual stresses, 37–9, 41–3, 57–9
Resistances
  buckling, 7, 59–60
  factors, 211–12, 225, 228, 231
  shear, 109, 112, 117
Restraining segments, 19–20
Restraints, purlin–roof panels, 171

Ribbed decks, composite beams, 119–20
Ridges, 164–5
Rigidities, 7–8, 47–8
Ritz procedure, 174, 176
Rolled sections, 7
Roofing, 161–5
Rotational restraints, 180–1

Safety
  checking, 214, 216
  factors, 81, 83–4, 85, 191–3
    need, 209–11
    selection, 215–16, 224–32
    web openings, 99
St Venant torsion, 143, 147, 157, 202–3
Salvadori's method, 18–20, 22
Second moment theory, 212
Sections, properties, 6–8, 24–5, 27, 47, 49, 51, 57, 83
Sensitivity studies, 234
Service ducts, 97, 118
Serviceability limit states, 234
Shallow beams, 15
Shanley's explanation, 46
Shape factors, 81, 84
Shear
  affecting web openings, 108
  beams, 163
  capacities, 98
  centre positions, 24, 47–8, 58, 59, 147, 153
  moduli, 41
  resistances, 109, 112, 117
  rigidity, 179–80
  yielding, 115
Shrinkages, 41
Simply supported beams, 5–8, 38, 49, 53, 61, 64–5, 79, 168, 190, 195, 204–5
Slender beams, 24, 29, 128
  elastic buckling, 75
Slenderness
  limited, 76, 78, 80, 81
  modified ratios, 57–61, 63, 65, 77–9, 85–6, 89
  ratios, 190
Span–depth ratios, 152
Spring constraint, beam columns, 171, 180
SSRC design guides, 187, 190, 198
Stability
  interaction
    criteria, 189–90, 192
    equations, 195–6
  lateral, 8
  length effect, 8
  matrix, 52
Statistical assessments, 86–7
Statistical data, 225, 229
Steel, mechanical properties, 41
Stepped beams, 27
Stiffness
  matrix, 52, 152, 155
  warping, 7–8
Stocky beams, 15, 76
Strain
  distributions, 44
  energies, 8–9, 141, 173
  hardening, 43–7, 64, 113, 116, 124
Strengths, increasing, 139
Stress
  analysis, 100–6
  concentration factors, 99
  distributions, 44, 109–11, 120–1
  resultant analysis, 109–28
  reversal, 111, 114
Stress–strain curves, 41
Sudden failure, 3, 5

T-beams, 153–4
Tangent modulus theory, 46–7, 76
Tapered beams, 26–7
Target reliability index values, 215, 227–9
Thin-walled beams, 137–59, 161–83
Thin webs, 15
Timoshenko analysis, 4
Torsional parameters, 8, 11, 16, 194, 196
Torsional restraint, 14–15, 180–1
Torsional rigidities, 7–8
Torsional–flexural theory, 166–9
Transfer matrix methods, 54, 55–6

Transverse loading, 10–11, 26, 61–2, 191
Twisting, 28–30
Two-span loading, 18, 64–6

UK design methods, 81–3, 90–1
Ultimate strengths
  analysis, 107–28
  equations, 189, 192
  reduced, 37
Uniform bending, 24–5, 86–7, 149, 153
  buckling, 5–8

Vierendeel analysis, 105, 106
Vinnakota calculations, 53–5
Von Mises yield criterion, 111

Wagner effect, 25
Web
  crippling, 179
  openings
    depth ratios, 97, 100, 103, 104, 107, 112
    elastic stress analysis, 100–7
Web—*contd.*
  openings—*contd.*
    geometries, 110, 117
    reason for, 97, 127
    reinforced, 122–5
    shapes, 98–9
    ultimate strength analysis, 107–28
    without reinforcement, 109–22
  yielding, 45
Web-post, deformation, 125–6
Weighting factors, 226–7, 229
Welded beams, 3–4, 38–9, 42–3, 58–9, 76–9, 87–8
Wind uplift loading, 168–79
Woolcock and Trahair method, 23, 24
Work done, 8

Yield stresses, 29, 41, 43, 85
  reduction, 88
Yielding, 43–5, 58, 59
  beam, 37, 64, 65
  criteria, 188–9, 191–2
  shear, 115

Z-beams, 161–83